T0181700

Progress in Geophysics

Series Editor

Alexander B. Rabinovich, P.P. Shirshov Institute of Oceanology, Russian Academy of Sciences, Moskva, Russia

The "Progress in Geophysics" book series seeks to publish a broad portfolio of scientific books, aiming at researchers, students in geophysics. The series includes peer-reviewed monographs, edited volumes, textbooks, and conference proceedings. It covers the entire research area including, but not limited to, applied geophysics, computational geophysics, electrical and electromagnetic geophysics, geodesy, geodynamics, geomagnetism, gravity, lithosphere research, paleomagnetism, planetology, tectonophysics, thermal geophysics, and seismology.

More information about this series at http://www.springer.com/series/15922

Victor Privalsky

Time Series Analysis in Climatology and Related Sciences

 Springer

Victor Privalsky
Water Problems Institute
Russian Academy of Sciences
Moscow, Russia

ISSN 2523-8388   ISSN 2523-8396   (electronic)
Progress in Geophysics
ISBN 978-3-030-58054-4   ISBN 978-3-030-58055-1   (eBook)
https://doi.org/10.1007/978-3-030-58055-1

This Springer imprint is published by the registered company Springer Nature Switzerland AG
The registered company address is: Gewerbestrasse 11, 6330 Cham, Switzerland

*To the memory of my daughter Maria
Privalsky and my teacher Akiva Yaglom*

Acknowledgements

The author is sincerely indebted to Alexander Benilov, Sergei Dobrovolski, M. Fortus, Max Malkin, Sergei Muzylev, Julie Rich, and Tatyana Vyruchalkina for their valuable help.

Special thanks are to Alexander Rabinovich for recommending this book to Springer.

Contents

Abbreviations

AAO	Antarctic Oscillation
AIC	Akaike information criterion
AGST	Annual global surface temperature
AMO	Atlantic Multidecadal Oscillation
AO	Arctic Oscillation
AR	Autoregressive
AR	Multivariate autoregressive
ARIMA	Integrated autoregressive and moving average
ARMA	Autoregressive and moving average
ARR	Autoregressive reconstruction
ASOI	Annual Southern Oscillation Index
AST	Annual surface temperature
BIC	Schwarz–Rissanen criterion
CAT	Parzen criterion of autoregressive transfer function
cph	cycles per hour
CRR	Correlation/regression reconstruction
DMI	Dipole Mode Index
ENSO	El Niño-Southern Oscillation
ESGF	Earth System Grid Federation
FRF	Frequency response function
FT	Fourier transform
GCM	General circulation model
GISS	Goddard Institute of Space Studies
GLOBE	Global annual surface temperature
GPCC	Global Precipitation Climatology Centre
hPa	Hectopascal
Im	Imaginary part of a complex-valued quantity
KWT	Kolmogorov–Wiener theory (of extrapolation)
LAND	Annual terrestrial surface temperature
LNH	Annual terrestrial surface temperature of the northern hemisphere

LSH	Annual terrestrial surface temperature of the southern hemisphere
MA	Moving average
MEI	Modoki El Niño Index
MEM	Maximum entropy method
MJO	Madden–Julian Oscillation
MTM	Multitaper method
NAO	North Atlantic Oscillation
NH	Annual surface temperature of the northern hemisphere
NINO	Oceanic component of ENSO
NPI	North Pacific Index
OCEAN	Annual sea surface temperature of the World Ocean
ONH	Annual sea surface temperature of the northern hemisphere
OSC	Order selection criterion
OSH	Annual sea surface temperature of the southern hemisphere
PDF	Probability density function
PDO	Pacific Decadal Oscillation
PDSI	Palmer Draught Severity Index
PNA	Pacific-North American Pattern
PQC	Prediction quality criterion
PSI	Hannan–Quinn criterion
QBO	Quasi-Biennial Oscillation
Re	Real part of a complex-valued quantity
RIC	Residual information criterion
RMM	Components of Madden–Julian Oscillation
RMS	Root mean square (value)
RMSE	Root mean square error
RPC	Relative predictability criterion
SH	Annual surface temperature of the southern hemisphere
SOI	Southern Oscillation Index
SSA	Singular spectrum analysis
SSN	Sunspot numbers
SST	Sea surface temperature
TPI	Tripole Index
TSI	Total solar irradiatiance

Chapter 1
Introduction

Abstract Most geophysical processes are random, and their analysis should be based upon theory of random processes and information theory. The main tool of analysis here is the autoregressive modeling of scalar and multivariate time series in time and frequency domains. After a brief description of theoretical basis, the scalar case continues from parametric and nonparametric analysis to the final stage—time series prediction. The text includes theory and examples of spectral estimation, description of extrapolation theory, and examples of its application in climatology, oceanography, and meteorology. Multivariate time series analysis is used for describing relations between scalar time series (teleconnections) and for time series reconstructions. The suggested solutions of both tasks disagree with the traditional approach and their advantages are demonstrated, in particular, by investigating a dependence of global surface temperature upon ENSO and reconstructing a simulated time series typical for climate data. A unique climatic process—Quasi-Biennial Oscillation—is analyzed in the frequency domain. Analyses of trivariate time series show the potential of multivariate autoregressive time and frequency domain approach. Time series approach is used for studying ice core, solar, and climate simulation data. The book contains recommendations that help to avoid erroneous steps in time series analysis.

Practically all geophysical processes as well as the processes occurring on the Sun are random, and studying them requires mathematical tools developed within theory of random processes and information theory. The time series obtained through observations over elements of the geophysical system atmosphere—ocean—land—cryosphere present the most reliable sources of information about the state of the system from the current time and back into the past at time scales from seconds (turbulence) to decades and even centuries (climate and related sciences including solar research). At climatic time scales, which are understood here as starting from years and going to anything longer, variations of the system's elements are random and should be described in a probabilistic mode. A successful prediction of climate—the most desirable and important goal of research in the area—cannot be achieved without the knowledge of probabilistic laws that control the system's behavior. This statement holds for both the physically based predictions through general circulation

© Springer Nature Switzerland AG 2021
V. Privalsky, *Time Series Analysis in Climatology and Related Sciences*,
Progress in Geophysics, https://doi.org/10.1007/978-3-030-58055-1_1

models and for statistical forecasting (prediction, extrapolation) of natural variations of climate and other geophysical processes.

When trying to move from having a climatic time series to predicting respective element of the system, one has to complete a number of tasks starting from preliminary processing and ending with a prediction equation complemented with a confidence interval for the predicted trajectory. This process includes the use of methods of time series analysis which can be classified for our purposes as methods for scalar (univariate) and multivariate (more than one scalar component) time series. In both cases, it will be necessary to study properties of the time series, first of all, whether it can be treated as stationary and then the probability density function, its statistical moments, and its probabilistic (traditionally called statistical) predictability. The theory of random processes and information theory became sufficiently developed for practical applications around the middle of the previous century and the methods developed by mathematicians since that time allow one to travel the road from preliminary processing to forecasting as the final step or, more often, to some stage preceding this final step. The forecasting must be done in accordance with theory of stationary random processes. These considerations are true for all geophysical and solar processes at any time scales with the exception of tidal phenomena controlled by the lunar and solar gravity forces.

With the scalar time series, the task is relatively simple: check the stationarity, define the probability density function, determine its statistical moments including its correlation function and spectral density, estimate statistical predictability, and forecast the time series if you need it and if the predictability is sufficiently high. The forecasting stage is the most important part of research, and it requires stochastic equations for obtaining a forecast and determining the error variance. This issue is discussed in the book for both scalar and bivariate cases.

The key stages include estimation of spectral density and statistical predictability, which is closely related to the spectral density. It is not obligatory in all situations to go through the forecasting stage, but the spectrum of the time series is undoubtedly the most important statistical characteristic of any stationary time series, and it should be estimated whenever a time series is being analyzed. The knowledge of time series spectrum is obligatory. This statement holds for multivariate time series as well, but then the quantity that is required for understanding and, possibly, predicting the time series is its spectral matrix.

In the case of multivariate time series which include several scalar components, the tasks of analysis and prediction are more complicated. The correlation and spectral density functions become matrices, which define the properties of the scalar components of the time series and relationships between them in time and frequency domains.

The time series analysis must be done in both time and frequency domains; its mathematical foundations are relatively complicated in the multivariate case, and examples of such analyses are difficult to find in Earth and solar sciences. Statistical prediction of time series is obtained as a function of its behavior in the past, and it is realized through the forecasting equation which is to be built for the time series at the analysis stage.

In climatology, solar science, and in some other areas, there is an additional task of reconstructing a time series on the basis of its dependence upon one or several other time series known on a longer time interval. Summing up, the tasks that must be solved in the process of time series reconstruction include

- autoregressive time domain modeling,
- estimation of frequency dependent functions that characterize the behavior of the time series and relations between its components,
- using the results of multivariate analysis in time and frequency domains for understanding interdependences between time series and for their reconstruction.

One of the goals of this book is to show that time series analysis applied to traditionally important tasks in climatology and related sciences within the framework of theory of random processes and information theory produces results that differ from what we see in literature. The goal is realized through describing the proper methods of analysis and demonstrating results of their application to scalar and multivariate (mostly, bivariate in this book) time series belonging to climatic processes and to other geophysical and solar data. A partial list of tasks discussed in the book includes time and frequency domains autoregressive modeling and spectral analysis of time series, their statistical forecasting, teleconnection research and analysis, and time series reconstruction.

The methods of analysis of scalar and multivariate time series are described in books written by mathematicians such as Bendat and Piersol (1966, 2010), Jenkins and Watts (1968), Box and Jenkins (1970), Reinsel (2003), Box et al. (2015), Shumway and Stoffer (2017) as well as by authors specializing in Earth sciences such as von Storch and Zwiers (1999) and Thomson and Emery (2014). Yet, many publications, especially in climatology but also in other sciences, disagree with theory of random process, information theory, and mathematical statistics. According to Thomson and Emery (2014, p. 425), even the mathematically proper methods of time series analysis are often misunderstood and used incorrectly. A key statistics of a scalar time series is the spectral density, but spectral estimation in climatology and related sciences suffers from the use of inadequate methods of spectral estimation, from the lack of confidence intervals for spectral estimates, and from application of a popular but erroneous test for statistical significance. This situation has led to discoveries of "statistically significant" spectral peaks at practically any frequency in the spectra of climatic and other time series even at time scales comparable to the time series length.

Statistical (more accurately, probabilistic, because mathematical statistics does not include theory of random processes and information theory) methods are often used to forecast geophysical processes, but the classical theory of extrapolation— one of the greatest achievements in theory of random processes—created by Andrey Kolmogorov and Norbert Wiener about 80 years ago remains essentially unknown in Earth and solar sciences. Leaving other issues aside, the theory deserves attention for the simple reason that it contains a mathematical proof of its ability to produce results with the smallest error variance in the class of linear methods of forecasting and the smallest possible error variance in the Gaussian case. In other words, if one has a Gaussian stationary time series, its prediction (forecasting, extrapolation) must be done in accordance with the Kolmogorov–Wiener theory of extrapolation (KWT). No other approach can lead to prediction with a smaller error variance. The disregard for the Kolmogorov–Wiener theory, which exists in geosciences, leads to results whose reliability is doubtful or nonexistent. Extrapolation methods based upon KWT ensure the smallest error variance of linear extrapolation irrespective of the time series probability density.

In climatology and related sciences, the multivariate analysis is applied mostly for finding teleconnections and for reconstructing past behavior of geophysical or solar processes. Traditionally, both tasks are treated through cross-correlation coefficients and regression equations, that is, within the framework of the classical mathematical statistics. However, regression analysis is not applicable to time series. More than 60 years ago, it has been shown in information theory that time series should not be treated as time-invariant random variables (Gelfand and Yaglon 1957); this means that the traditional approach cannot produce correct solutions of those two tasks. The proper methods of multivariate time series analysis have been developed at least half a century ago. Seemingly, the study of surface temperature in North Atlantic by this author (Privalsky 1988) was the first example of bivariate autoregressive time and frequency domain analysis in Earth and solar sciences. The paper by Privalsky and Jensen (1995) and the book by Thomson and Emery (2014, p. 433) contain direct warnings against applying the cross-correlation coefficient and regression equation to time series. A similar comment regarding the role of serial correlation is given in von Storch and Zwiers (1999, p. 157). The results obtained with the traditional approach cannot be trusted.

The book shows how to analyze, reconstruct, and forecast scalar and multivariate time series in agreement with theory of random processes, information theory, and mathematical statistics. In each case, a brief description of theoretical basis is accompanied with examples, which hopefully are detailed enough to allow the reader to learn and apply methods of time series analysis and prediction still practically unknown in climatology and other Earth and solar sciences. Most of the software necessary for mathematically proper analysis of scalar and multivariate time series is available in free and commercial software packages.

The book contains fifteen chapters. The chapters from 2 to 6, which deal with the scalar time series, are mostly dedicated to the methods of spectral analysis and to results that can be obtained by applying the proper methods of analysis and prediction to geophysical time series.

Chapter 2 contains general information from probability theory and theory of random processes, which is necessary for understanding how to analyze time series. It also contains examples of time series models common in climatology and geophysics in general. (The understanding of geophysics in this book does not include the solid body of the Earth.)

Chapter 3 describes two types of spectral analysis: the nonparametric, when the spectrum is estimated directly from the time series or from its estimated covariance function, and parametric, which begins with stochastic modeling of the time series in the time domain and then uses the time domain model to calculate the spectral estimate. A key point in the parametric—autoregressive in this book—spectral analysis is choosing the order of the model. This is an issue of prime importance, and the examples given in the book for both scalar and multivariate time series allow the reader to understand the necessity to correctly determine the model's order for obtaining reliable estimates of time series characteristics in both time and frequency domains.

Chapter 4 contains a brief description of practical time series analysis including the role of the sampling interval, trend analysis, testing the hypotheses of stationarity and ergodicity, and the role of filtering—the technique which is hardly necessary in most cases and especially in the case of parametric analysis. The ability of autoregressive spectral analysis to detect periodic components and correctly determine their frequencies is demonstrated with the time series of sea level observations of length of 10^5 h and the autoregressive order 10^4. The process of time and frequency domain analysis is illustrated with an example using the time series of sunspot numbers.

Chapter 5 is essentially an attempt to describe statistical properties common for climatic time series. In addition to already known results, a number of time series of climate indices and long-term oscillations have been analyzed by using both the autoregressive (maximum entropy) modeling and the nonparametric Thomson's multitaper method to verify the hypothesis of climate as a Markov process. The Markov hypothesis is confirmed to be quite common with the exception of climate data averaged over large parts of the globe. The chapter also contains examples of analysis of several rather unique geophysical time series: the Quasi-Biennial Oscillation (QBO), Madden–Julian Oscillation (MJO), and atmospheric and oceanic components of El Niño—Southern Oscillation (ENSO).

Chapter 6 is dedicated to the theory, the autoregressive method, and examples of extrapolation of stationary time series within the Kolmogorov–Wiener theory. The examples given in the chapter illustrate KWT-based forecasts of four geophysical processes: the annual global surface temperature, the monthly values of QBO, the oceanic component of the ENSO phenomenon, and the Madden–Julian Oscillation. In particular, it shows that the ENSO's atmospheric component—the Southern Oscillation Index (SOI)—is practically unpredictable with statistical methods.

The results of extrapolation obtained through methods not related to KWT must be compared with the results following from KWT. Using the Monte Carlo approach for determining the variance of forecast error cannot be regarded as a proof of the forecast efficiency. The KWT approach provides mathematically correct estimates of forecast error variance, and if the variance is large, it cannot be regarded as a drawback of the theory; it just means that the stationary random process which is being forecasted has low predictability.

A brief description of bivariate time series analysis is given in Chap. 7. The mathematical quantities lying in the basis of analysis—bivariate stochastic difference equations, innovation covariance matrices, and spectral matrices—are presented there along with the coherence function, coherent spectrum, and gain and phase factors. The stochastic difference equations and functions of frequency provide a thorough description of the stationary random process which generated the time series. The latter statement is less comprehensive if the process is not Gaussian, but the Gaussian

6

1 Introduction

distribution is often acceptable for climatic and other geophysical data. Relating time series to each other through the cross-correlation coefficient and regression equation is not correct because it can be done only to random variables; those variables do not depend upon time and frequency, and they do not have a correlation function or spectral density.

The Granger causality concept closely related to the Kolmogorov–Wiener theory of extrapolation and to the coherence function that describes interdependence between the components of bivariate time series is discussed as it is applied to analysis of time series in climatology and related disciplines. A simple and seemingly original technique is suggested for determining contributions of time series behavior in the past to its variance and to the variance of other components of multivariate time series.

Chapter 8 deals with application of time and frequency domain analysis of bivariate time series to teleconnection research; it contains examples of studying interdependence between the ENSO components and the effect of ENSO's oceanic component upon the annual surface temperature of the globe and its major parts. At climatic time scales, the ENSO system is shown to be close to a bivariate white noise with a high coherence between SOI and sea surface temperature in equatorial Pacific (NINO). In spite of the low cross-correlation between NINO and the surface temperature of the globe and its eight major parts, the ENSO's oceanic component NINO is shown to form strong teleconnection systems with the nine spatially averaged annual surface temperature sets; the coherence functions between NINO and surface temperature go up to 0.9 at ENSO's natural frequency of about 0.2 cycles per year (cpy). Further analysis shows that the contribution of NINO to surface temperature is minor because the spectral density of all nine time series of annual surface temperature at that frequency is small. Chapter 8 also contains probably the first in the Earth and solar sciences example of multivariate predictability analysis within the Kolmogorov–Wiener theory using as an example the bivariate time series of Madden–Julian Oscillation.

Chapter 9 is dedicated to time series reconstruction. It begins with some comments on the history of reconstruction task in climatology and related sciences and continues with a critical evaluation of the traditional method based upon the regression equation and cross-correlation coefficient. The earliest critical comments about the method had been made by its author A. Douglass but remained unheeded for many decades. The proposed autoregressive reconstruction method lies within the framework of theory of stationary random processes and information theory; it allows one to estimate properties of the target/proxy time series in time and frequency domains, reconstruct the target through the time domain autoregressive model, and verify the efficiency of reconstruction through frequency domain analysis. The efficiency of the method is confirmed through reconstruction of a simulated bivariate time series with the reconstruction target known over the entire interval of proxy observations.

In Chap. 10, the autoregressive spectral analysis is applied (seemingly, for the first time) to study the behavior of QBO as a function of altitude and frequency using six bivariate linear systems in which the time series obtained at the 10 hPa level is the input, and the six time series at levels from 15 hPa to 70 hPa are the outputs. The frequency of QBO at 0.43 cpy is shown to be very stable within the entire stratospheric layer between 31 km and 18 km of altitude, and the coherence between oscillations at the uppermost and lower levels is extremely high reaching 0.95–0.99 within the entire set. The rate of downward propagation of QBO is shown to become slower as the system moves to lower levels. The time domain analysis of the QBO system through the technique suggested in Chap. 7 allows one to detect the presence of upward flows having the time scales of months and years.

Simulations of surface temperature and ENSO components within the IPCC's Coupled Model Intercomparison Project with 47 general circulation models are compared in Chap. 11 with respective data of observations at climatic time scales. The experiments include the ENSO system, its effect upon surface temperature averaged over the globe and its major parts, and the models' ability to reproduce surface temperature within the continental USA. The models were found to possess some positive features such as the ability to properly simulate the nonmonotonic spectra of ENSO components, their dependence upon each other, and basic statistics of surface temperature. However, the simulation results have a strong variability from model to model and, what is quite important, the models incorrectly ascribe a significant contribution of ENSO to spatially averaged surface temperature at low frequencies. Chapter 11 is written in co-authorship with V. Yushkov.

An attempt to introduce methods of scalar and bivariate time series analysis into paleoclimatology is taken in Chap. 12 by using the results obtained from analysis of eight time series of ice core data in Greenland and Antarctica. It describes the procedure of analysis of proxy data in time and frequency domain and shows that the persistence of those proxy time series which are used in climatology to reconstruct climate can be quite different geographically. Bivariate analysis of ice core data shows that they are not statistically related to each other and, consequently, do not carry a common climatic "signal" even when the distance between the sites amounts to less than 300 km (Greenland) or 600 km (Antarctica).

Chapter 13 is dedicated to studying statistical properties of sunspot numbers (SSN) and total solar irradiance (TSI) including their statistical predictability. The time series of SSN and TSI can be regarded as stationary, their spectra contain a peak at the solar cycle frequency, and their statistical predictability is relatively high. However, the prediction quality diminishes at a monthly sampling rate because the solar cycle is asymmetric and the probability density functions of ice core data strongly differ from Gaussian. Formally, examples of forecasts are correct, but it is obvious that the task of SSN and TSI extrapolation needs to be studied through a nonlinear approach.

Chapter 14 contains a description of time series analysis for the case of a simulated trivariate linear system with two inputs and one output. The example shows what can be learned about a multivariate time series when it is analyzed parametrically in time and frequency domains. It includes a multivariate stochastic difference equation, spectra, multiple, and partial coherence functions and partial coherent spectra as well as gain and phase factors for each tract of the system. The multivariate autoregressive spectral analysis is shown to be a powerful tool, but it does not seem to have ever been applied in Earth or solar sciences. The method described in the chapter is used for analysis of a trivariate time series consisting of data that describe the behavior of global and other spatially averaged time series.

The concluding Chap. 15 presents the author's view of the current situation with time series analysis in climatology and related sciences; it also contains recommendations regarding the practices which should be avoided in analysis of climatic and other similar time series.

The mathematical methods used throughout the book, with one or two minor exceptions, are not new. They have been used for time series analysis since at least half a century ago and shown to be reliable in many applied disciplines, especially in engineering. An essential step taken in this book is combining the autoregressive time domain approach with autoregressive spectral analysis of multivariate time series. This has been done long ago for the scalar case (Burg 1967), but simultaneous parametric time and frequency domain multivariate analysis seems to be practically unknown in Earth and solar sciences.

References

Bendat J, Piersol A (1966) Measurement and analysis of random data. Wiley, New York
Bendat J, Piersol A (2010) Random data. Analysis and measurements procedures, 4th edn. Wiley, Hoboken
Box GEP, Jenkins GM (1970) Time series analysis. Forecasting and control. Wiley, Hoboken
Box G, Jenkins M, Reinsel G, Liung G (2015) Time series analysis. Forecasting and control, 5th edn. Wiley, Hoboken
Burg J (1967) Maximum entropy spectral analysis. Paper presented at the 37th Meeting of Society of Exploration Geophysicists, Oklahoma City, OK, October 31, 5 pp
Gelfand I, Yaglom A (1957) Calculation of the amount of information about a random function contained in another such function, Uspekhi Matematicheskikh Nauk, 12:3–52, English translation: American Mathematical Society Translation Series 2(12):199–246, 1959
Jenkins G, Watts D (1968) Spectral analysis and its applications. Holden-Day, San Francisco
Privalsky V (1988) Stochastic models and spectra of interannual variability of mean annual sea surface temperature in the North Atlantic. Dynam Atmos Ocean 12:1–18
Privalsky V, Jensen D (1995) Assessment of the influence of ENSO on annual global air temperature. Dynam Atmos Ocean 22:161–178
Reinsel G (2003) Elements of multivariate time series analysis, 3rd edn. Springer, New York
Shumway R, Stoffer D (2017) Time series analysis and its applications, 4th edn. Springer, Heidelberg
Thomson R, Emery W (2014) Data analysis methods in physical oceanography, 3rd edn. Elsevier, Amsterdam
von Storch H, Zwiers F (1999) Statistical analysis in climate research, 2nd edn. Cambridge University Press, Cambridge

Chapter 2
Basics of Scalar Random Processes

Abstract The chapter contains elements of theory of random processes required for time series analysis and for understanding its results. The time series discussed in this book belong to the class of stationary random processes, which means that their statistical properties averaged over an ensemble of realizations of the process do not depend upon the time origin. If averaging over an ensemble of realizations is equivalent to averaging over time of any single realization, the process is ergodic. If a stationary process is Gaussian, it is also ergodic. In our research, we mostly have just one sample realization (one time series) of finite length and by extending the properties estimated from a single time series to the entire process under the study, we assume that the process is ergodic. The most important characteristics of any time series include the covariance function and spectral density, which describe the time series properties in the time and frequency domains, respectively. The two stationary discrete random processes typical for climatology and other disciplines are white noise and Markov chains. Examples of the true and estimated correlation functions and spectral densities including their sampling variability are briefly discussed for simulated time series of length $N = 100$.

The type of data studied in this book is time series, that is, sequences of numerical quantities distributed in time (Wiener 1948). Time series analysis is a part of theory of random processes—a mathematical discipline. Therefore, the issue of definitions plays a key role in applications of time series analysis. A more detailed definition given by N. Wiener in his second classical book on extrapolation, interpolation, and smoothing of stationary time series is more detailed: "Time series are sequences, discrete or continuous, of quantitative data assigned to specific moments of time and studied with respect to the statistics of their distribution in time" (Wiener 1949). According to A. Yaglom, the terms time series and random function depending upon "mostly but not always time" (Yaglom 1962, 1986) are equivalent. Thus, the dependence upon time is an inherent property of time series, and it makes them cardinally different from time-invariant random variables, which belong to the classical mathematical statistics. This statement holds both for scalar and multivariate time series.

© Springer Nature Switzerland AG 2021
V. Privalsky, *Time Series Analysis in Climatology and Related Sciences*,
Progress in Geophysics, https://doi.org/10.1007/978-3-030-58055-1_2

The time series and random functions of time in this book always consist of discrete random variables and are regarded as sample records of discrete random processes. In contrast to time-invariant sets of random variables, statistical properties of time series will change if the order of its terms is changed. The time series in chapters from 2 through 6 are univariate, or scalar.

The concept of time series, or random functions, does not exist in the classical mathematical statistics, and time series analysis is based upon theory of random processes; consequently, it requires special mathematical tools (Yaglom 1962). Some of those tools are described in this book and are applied to analyze and forecast time series which are studied in Earth and solar sciences.

2.1 Basic Statistical Characteristics

The notation used in this book for time series is x_t, $t = \Delta t, \ldots, N\Delta t$, where N is the number of terms in the time series and Δt is the time interval between consecutive terms. The sampling rate is the number of samples per unit time and, as a rule, the sampling rate here is one sample per year or one sample per month, that is, $\Delta t = 1$ year or $\Delta t = 1$ month. The notation for the time series can also be given as x_t, $t = 1, \ldots, N$ having in mind that $\Delta t = 1$. In what follows, the time series is considered as long if its length exceeds the largest time scale of interest by orders of magnitude. Otherwise, the time series has to be treated as short.

A major characteristic of a time series is its probability density function $p(x)$, which defines the probability of encountering different numerical values of x_t. The common abbreviation for $p(x)$ is PDF. The PDF of a scalar time series is characterized with its statistical moments, or statistics, such as mean value and variance (central moments), covariance function and spectral density (mixed moments), and with higher central moments such as skewness and kurtosis.

There are many different types of PDFs, but the function which is most important for practical time series analysis (if its application can be justified for a given time series) is the Gaussian or normal, probability density function

$$p(x_t) = \frac{1}{\sigma_x \sqrt{2\pi}} \exp[-(x_t - \bar{x})^2/2\sigma_x^2], \qquad (2.1)$$

where the mean value

$$\bar{x} = \lim_{N \to \infty} \frac{1}{N} \sum_{t=1}^{N} x_t \qquad (2.2)$$

and

$$\sigma_x^2 = \lim_{N \to \infty} \frac{1}{N} \sum_{t=1}^{N} (x_t - \bar{x})^2 \tag{2.3}$$

is the variance of the time series. The positive square root σ_x of the variance is called the root mean square (RMS) value or standard deviation. The variance (and RMS) describes the variability of the process: a larger variance means a larger dynamic range of numerical values of the process. These definitions are true for random functions of time and for sequences of time-invariant random variables. The Gaussian PDF of random variables is completely described with the mean value $\bar{x}$ and the variance σ_x^2. By default, it is generally assumed here that time series are generated by Gaussian (normally distributed) random processes. This assumption does not limit or degrades the properties and abilities of the methods used in this book for time series analysis because the Gaussian probability distribution means the best possible results in all cases when the method is linear. The analysis of non-Gaussian time series requires estimation of the same statistical characteristics as in the Gaussian case and then some higher statistical moments. The methods applied here are linear, and they cover all traditional tasks of time series analysis, including spectral estimation and statistical (probabilistic) forecasting. The normal hypothesis should always be tested for the actual time series that is being analyzed.

The skewness and kurtosis can be helpful for analyzing geophysical and solar time series as measures of PDF's asymmetry and tail properties, respectively. They are mostly used for determining whether the PDF is close enough to a Gaussian (normal) distribution. Specifically, if the absolute values of standardized skewness and standardized kurtosis do not exceed 2, the time series can be generally regarded as Gaussian. (Standardized skewness and kurtosis are found by dividing respective estimate by $\sqrt{6/N}$ and by $\sqrt{24/N}$).

The PDFs and their central statistical moments completely describe properties of sets of random variables, which do not depend upon the time argument. In contrast to random vectors, time series present samples of random processes and their description is not possible without mixed statistical moments such as covariance and correlation functions in the time domain and the spectral density in the frequency domain. These functions are defined in Sect. 2.4.

2.2 Deterministic Process

In contrast to time series consisting of random variables, one can imagine a time-dependent process that follows a specific mathematical law, which excludes any randomness. Respective data present a deterministic process. Deterministic phenomena or processes can be predicted at any lead time without an error.

At climatic time scales (longer than one year), the only geophysical processes that can be regarded as deterministic are those that are caused by astronomical factors. With one exception (a barely detectable tidal harmonic with period of 18.61 years), the time scales of such processes extend to millennia and longer (e.g., Monin 1986) and their effects upon climate at the practically important time scales of year-to-year variability, decades and centuries are negligibly small. At smaller time scales, the only deterministic process in the Earth system is tides, which can be predicted practically precisely in the open ocean and along most shorelines. With the exception of solar and lunar tides, all other geophysical and relevant solar processes are random. Moreover, in some coastal areas (e.g., Newlyn, UK), the sea level variations can be better predicted in a different manner, "without astronomical prejudice and fully allowing for the presence of noise" (Munk and Cartwright 1966). An earlier method of tidal prediction without tidal harmonics had been developed by A. Duvanin (1960).

2.3 Random Process

A random (stochastic, probabilistic) process is any process running in time and controlled by probabilistic laws (Doob 1953). A random process consists of an infinite set of all possible sample records (time series, random functions of time) generated supposedly as the results of repeated experiments or observations. The random processes discussed here are always discreet, that is, the time argument takes only discrete values.

There are two major types of random processes: stationary and nonstationary. A stationary process is the process whose statistical properties determined by averaging at different time origins over the infinite set of sample records of the process do not depend upon the time origin. A nonstationary process does not possess this property. The time series studied in this book belong to the class of stationary random processes.

Consider this concept using the mean value and variance as examples. Let $x_{i,t}$, $i = 1, \ldots, M$; $t = 1, \ldots, N$ be an ensemble of sample records of a random process as shown in Fig. 2.1.

The mean value (or mathematical expectation) $\hat{x}(t)$ at time t is defined as the limit of the sum of M values of $x_{i,t}$ at time t divided by the number of sample records M as M tends to infinity:

$$\hat{x}(t) = \lim_{M \to \infty} \frac{1}{M} \sum_{i=1}^{M} x_{i,t}. \tag{2.4}$$

A random process is stationary with respect to the mean value if the mean values $\hat{x}(t)$ are statistically the same for all t.

If it is also true that the variance

$$\hat{\sigma}_x^2(t) = \lim_{M \to \infty} \frac{1}{M} \sum_{i=1}^{M} [x_{i,t} - \hat{x}(t)]^2 \tag{2.5}$$

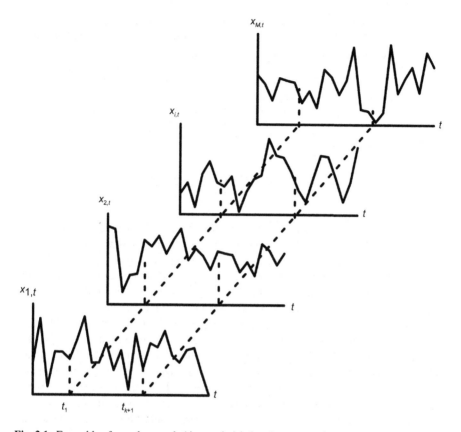

Fig. 2.1 Ensemble of sample records (time series) belonging to a random process

of the process possesses the same property, that is, it takes statistically the same value for all t, then we have a second order stationary random process.

According to the above definition, one sample record (time series) is not sufficient for studying a random process. This statement is true even if the sample record is known on the entire time axis from $-\infty$ to $+\infty$. Therefore, studying a random process requires a set of sample records (theoretically, an infinite set), so that all characteristics of the process (i.e., its PDF and its statistical moments) are determined by averaging over the ensemble of sample records. This cannot happen in geophysics in general, in climatology, or in solar physics where many if not all time series are unique. Therefore, by studying a single sample record of a stationary random process and extending the results of analysis to the entire process, one assumes by default that the results obtained by analyzing its properties through averaging over time coincide with the results of averaging over an ensemble of such records at different time origins. A random process that possesses this property is called ergodic. An ergodic process is always stationary.

Statistical characteristics of a random process can be determined by averaging over the time index t for every realization $x_{i,t}$ rather than over the ensemble of time series $x_{i,t}$ at every value of i. Then, the mean value is

$$\bar{x}(i) = \lim_{N \to \infty} \frac{1}{N} \sum_{t=1}^{N} x_{i,t} \qquad (2.6)$$

and the variance

$$\bar{\sigma}_x^2(i) = \lim_{N \to \infty} \frac{1}{N} \sum_{t=1}^{N} [x_{i,t} - \bar{x}(i)]^2. \qquad (2.7)$$

If the mean values $\bar{x}(i)$ and variances $\bar{\sigma}_x^2(i)$ estimated by averaging any sample record $x_{i,t}$ over time $t = 1, 2, \ldots$ coincide with the mean values $\hat{x}(t)$ and variances $\hat{\sigma}_x^2(t)$ estimated by averaging over the ensemble of sample records $x_{i,t}$, $i = 1, 2, \ldots$, then we are dealing with a second order ergodic process. In this case, the notations for the mean value and variance can be written as $\bar{x}$ and σ_x^2. In what follows, we will drop the dependence of the time series upon the sample record number.

The process is ergodic if the same property holds for its other statistical characteristics. One can say that the process is ergodic if the results of its analysis by averaging over the ensemble of its sample records (time series) statistically coincide with the results of analysis by averaging over time at any time origin. A nonstationary process cannot be ergodic. The expression "statistically the same value" or "statistically coincide" means that the estimates obtained for different sample records of finite lengths differ from each other not more than could be expected due to the sampling variability of estimates (variability from sample to sample). This means that the estimates stay within the range that can be expected due to the differences between estimates obtained by analyzing different finite samples (or sample records) of the same process.

When the properties of a process are being studied by analyzing a single sample record, it means an implicit assumption that the time series has been generated by an ergodic random process, which, of course, is a very serious statement. However, the assumption of ergodicity is correct if a stationary random process is Gaussian, that is, if its probability density function follows the Gaussian law given with Eq. (2.1). According to Bendat and Piersol (2010, p. 11), "in practice, random data representing stationary physical phenomena are generally ergodic". Moreover, a non-ergodic stationary random process can always be regarded as a mixture of ergodic components (Yaglom 1986, p. 217). If, on the contrary, one is studying how statistical properties of a time series change in time and finds that the differences between time-dependent estimates of statistical characteristics are significant according to some properly selected criteria, then the random process is nonstationary by definition; consequently, it is not ergodic and its statistical properties cannot be learned by analyzing a single time series. An example of such studies would be the wavelet

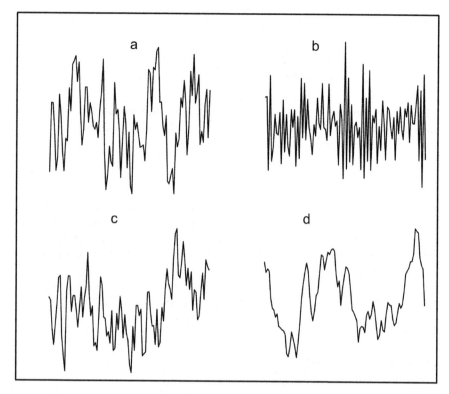

Fig. 2.2 Simulated examples of time series: **a** river streamflow, **b** not a geophysical process, **c** sea surface temperature, **d** water level of terminal lake

analysis applied to a time series which represents a random process with the purpose of detecting whether the process is stationary. A nonstationary time series can still be analyzed, but it has to be very long. In particular, Bendat and Piersol (2010) discuss analysis of nonstationary time series in Chap. 12 of their remarkable book.

Examples of some processes more or less common for variations of climate or other geophysical processes are shown in Fig. 2.2. The example b) is a sample of a random process whose energy increases as the time scales of its variations become smaller.

2.4 Covariance and Correlation Functions

In contrast to random variables, the random functions (time series) are time-dependent and, therefore, their properties in the time domain need to be described with statistical moments that reflect the dependence of the time series upon time. Such a mixed statistical moment obtained by averaging products of time series values separated from each other by k time intervals is the covariance function $R(k)$:

$$R(k) = \lim_{N \to \infty} \frac{1}{(N-k)} \sum_{t=1}^{N-k} (x_t - \bar{x})(x_{t+k} - \bar{x}), k = 1, 2, \ldots, K. \qquad (2.8)$$

Divided by the time series variance σ_x^2, it produces the correlation function $r(k)$ that presents a sequence of correlation coefficients between values of the time series x_t and x_{t+k} separated by k time intervals:

$$r(k) = R(k)/\sigma_x^2. \qquad (2.9)$$

The covariance and correlation functions are even functions of argument k so that $R(k) = R(-k)$ and $r(k) = r(-k)$. The covariance function's dimension is the square of the time series' dimension; if the time series is measured in Kelvin (K), the covariance function's dimension is K^2. The correlation function is dimensionless.

In what follows, it will be assumed that the mean values of all time series in this book are equal to zero $(\bar{x} = 0)$. The exceptions to this rule include Eqs. (4.1) and (4.2), the test for stationarity in Chap. 4, and examples of extrapolation in Chap. 6.

2.5   Spectral Density

The most essential statistical characteristic of a scalar time series is its spectral density $s(f)$; it defines how the energy of time series variations changes over the frequency f (the number of cycles per unit time), that is, how it varies in the frequency domain. There are different ways to define the spectral density (also called the spectrum), and one of them is to present the spectral density of a stationary random process as a Fourier transform of the covariance function:

$$s(f) = \sum_{k=-\infty}^{\infty} R(k)e^{-i2\pi kf \Delta t}, \qquad (2.10)$$

where $i = \sqrt{-1}$ and f is the cyclic frequency defined from $f = 0$ through the Nyquist frequency $f_N = 1/2\Delta t$. If $\Delta t = 1$ year, the frequency is measured in cycles per year (cpy) or year^{-1}.

The covariance function can be presented as the inverse Fourier transform of the spectral density:

$$R(k) = \int_{-f_N}^{f_N} s(f)e^{i2\pi kf \Delta t} df. \qquad (2.11)$$

According to Eq. (2.11), the spectral density dimension is the square of the time series dimension divided by frequency. Thus, if x_t is measured in millimeters and frequency in cycles per year, the dimension of the spectral density is mm^2/cpy, or mm$^2 \times$ year. Equations (2.10) and (2.11) constitute the Wiener–Khinchin theorem for discrete random processes. The argument $k = 0, \pm 1, \ldots$ of the covariance function is discrete, while the spectral density is a continuous function of frequency f. The covariance and correlation functions as well as the spectral density do not exist for sets of random variables (random vectors) because random variables do not depend upon time and, consequently, upon frequency.

An important special case of random processes is the white noise: a sequence a_t of identically distributed and mutually independent random variables. The white noise concept allows one to introduce the class of linearly regular, or regular, random process, which is defined as the process at the output of a linear system (linear filter) with a white noise at the input (the Wold decomposition):

$$x_t = \sum_{j=0}^{\infty} \psi_j a_{t-j}, \tag{2.12}$$

where ψ_j are filter's coefficients. Normally, it is assumed that $\psi_0 = 1$. Thus, a regular random process presents a linear transformation of a white noise. The upper limit in the sum in Eq. (2.12) can be finite while the lower limit is zero because if it is less than zero the process will be unrealizable physically. If the coefficients ψ_j do not change with time and if $\sum_{j=0}^{\infty} \left| \psi_j^2 \right| < \infty$, the output x_t of the filter belongs to a stationary random process. The quantity a_t is also called the innovation sequence. The mean value $\bar{a}$ of the white noise in Eq. (2.12) is always zero, so that the mean value of the process x_t is zero as well. All time series in this book, with the exception of tides, belong to regular stationary processes.

The spectrum $s(f)$ of a regular random process is an absolutely continuous function of frequency f, which means, in particular, that a regular random process cannot be presented as a finite or countable infinite set of periodic functions. In other words, a regular random process does not contain any harmonics. If one assumes the presence of harmonic oscillations in the process, the process loses the property of linear regularity, which may have negative consequences for its analysis and forecasting. However, the process stays regular if its spectrum contains sharp peaks that take an arbitrarily narrow but finite interval of frequencies. An example will be given in Chap. 4.

The correlation function of white noise is

$$r(k) = \begin{cases} 1, & k = 0 \\ 0, & k \neq 0 \end{cases} \tag{2.13}$$

and its spectral density is a constant:

$$s(f) = 2\sigma_a^2 \Delta t, 0 \le f \le 1/2\Delta t \tag{2.14}$$

The ideal white noise with discrete time parameter is not physically realizable, and results of its simulations are sequences of pseudo-random numbers. (A continuous white noise is not realizable as well because its variance must be infinite.)

The book containing more detailed information about basics of theory of random processes at the engineering level is Bendat and Piersol (2010, Chaps. 1 through 5).

The length of all our time series is finite; therefore, all statistical characteristics calculated from them contain sampling or estimation errors which must be evaluated to ensure reliability of results of analysis. Briefly, if φ is the usually unknown true value of a quantity that is being estimated and $\hat{\varphi}$ is an estimate of φ obtained from the time series, then the error of the estimate is characterized with the mean square error $E[(\hat{\varphi} - \varphi)^2]$ where E is the symbol of mathematical expectation. The mean square error includes the bias and random errors and the customary estimate of the estimate's accuracy is the root means square error

$$\text{RMSE} = \sqrt{\sigma^2[\hat{\varphi}] + b^2[\hat{\varphi}]} \tag{2.15}$$

where the random error

$$\sigma[\hat{\varphi}] = \sqrt{E[\hat{\varphi}^2] - E^2[\hat{\varphi}]}$$

and the bias (or the systematic error)

$$b[\hat{\varphi}] = E[\hat{\varphi}] - \varphi.$$

The ratios of the random and systematical errors to the true value of the parameter φ

$$\varepsilon_r = \frac{\sqrt{E[\hat{\varphi}^2] - E^2[\hat{\varphi}]}}{\varphi}$$

and

$$\varepsilon_b = \frac{b[\hat{\varphi}]}{\varphi} = \frac{E[\hat{\varphi}]}{\varphi} - 1$$

are called normalized random and bias errors so that the total normalized error of estimation

$$\varepsilon = \frac{\sqrt{\sigma^2[\hat{\varphi}] + b^2[\hat{\varphi}]}}{\varphi} = \frac{\sqrt{E[(\hat{\varphi} - \varphi)^2]}}{\varphi}.$$

This quantity defines the confidence interval for a single estimate of φ

$$\left[\frac{\hat{\varphi}}{1+\alpha\varepsilon} \le \varphi \le \frac{\hat{\varphi}}{1-\alpha\varepsilon}\right],$$

where α is a coefficient depending upon the probability density function that describes the estimate's error. Assuming that the error has a Gaussian distribution, the coefficient α equals approximately 1.6 and 2 for confidence levels 0.9 and 0.95, respectively. The last two values show the probability that the confidence interval will cover the true value of φ. The quantity φ can be a parameter (e.g., a time series variance) or a function such as a covariance or correlation function, a spectral density, or any other function of time or frequency.

If the mathematical expectation of an estimate $\hat{\varphi}$ coincides with the true value φ, the estimate is called unbiased. It is an important property, and practically, it means that the estimated value comes closer to the true one as the time series length increases. If the method of estimation ensures that the mean squared value of the estimate error is minimal as compared with other methods, the estimate is called efficient. If the probability of estimate's deviation from the true value of the parameter tends to zero as the length of the time series increases, the estimate is consistent. A reliable estimate should possess all these properties.

It should always be remembered that any estimate of any parameter or function is absolutely useless if it is given without information about its reliability meaning, in particular, a confidence interval for the estimate.

2.6   Examples of Geophysical Time Series and Their Statistics

Consider several characteristic examples of climatic or other geophysical processes represented here with short ($N = 100$) simulated dimensionless time series.

The random processes of a white noise type (Fig. 2.3) are quite common and may include atmospheric pressure, precipitation, and climate indices, for example, the North Atlantic Oscillation. As seen from the figure, the estimates obtained from sample records of white noise do not coincide with the true values of the correlation function and spectral density. This happens due to the sampling variability phenomenon, which occurs whenever the length of the time series that is being analyzed is finite, that is, always. The confidence intervals for the estimated functions in this and other figures in this chapter are not shown intentionally to illustrate the phenomenon of sampling variability.

Another process often encountered in climate and geophysical data in general is the Markov chain, that is, a discrete random process whose future state is independent of its past under the condition that the state of the process at the current time is known. The correlation function of a Gaussian Markov process is defined by its first value $r_1 \equiv r(1)$:

$$r(k) = r_1^{|k|} \tag{2.15}$$

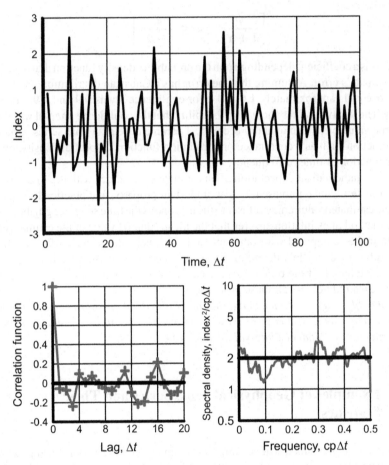

Fig. 2.3 White noise: a sample record, true (black) and estimated (gray) correlation function and spectral density

The correlation function and spectrum of a Markov process with a positive coefficient r_1 decrease monotonically as shown in Fig. 2.4. This Markov model is common for many climatic time series. Typical examples are the annual river streamflow and sea level variations. In 1976, K. Hasselmann suggested that the behavior of the climate system can be described with a Markov process (also see Dobrovolski 2000).

A Markov model is often used in climatology to determine statistical reliability of peaks in spectral estimates calculated from time series of geophysical observations. This approach is erroneous because it can only show whether the time series can be regarded as belonging to a Markov process.

Theoretically, the value of r_1 can be negative; then, the correlation function will be changing its sign at each lag and the spectrum will be growing with frequency. However, such processes (example b in Fig. 2.2) do not seem to exist in geophysical phenomena.

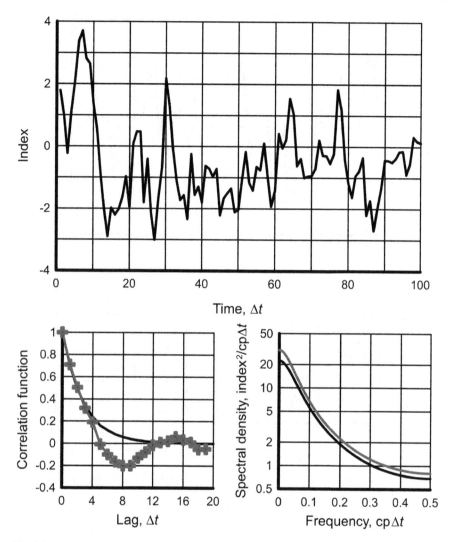

Fig. 2.4 Markov chain: a sample record, true (black) and estimated (gray) correlation function and spectral density

The immediate future behavior of a second-order Markov chain depends only upon its current state at time t and upon its state at time $t - 1$. Examples with simulated data are given in Figs. 2.5, 2.6 and 2.7. This type of time series is common for some climate indices such as the Atlantic Multidecadal Oscillation, water levels of terminal lakes, and components of the El Niño-Southern Oscillation (ENSO) system. As seen from Fig. 2.5, the true correlation function and spectral density may diminish monotonically and the dynamic range of the spectral density remains rather small (close to one order of magnitude). The estimates generally do not deviate very much

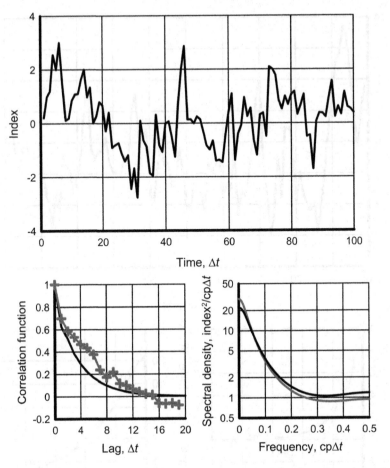

Fig. 2.5 Markov chain of order 2: a sample record, true (black) and estimated (gray) correlation function and spectral density

from the true quantities, which shows that in some cases a short time series can be sufficient for a reliable estimation of its stochastic model.

The situation becomes different with the annual water level variations in terminal lakes, such as the Caspian Sea, Great Salt Lake, or Lake Balkhash (Fig. 2.6). The terminal lake system presents a rather unique phenomenon that allows one to obtain a physically based dynamic stochastic system (e.g., Privalsky 1988). Water level variations in terminal lakes are slow, with sequential values heavily dependent upon the history of the process. This property results in a slowly changing correlation function and a quickly decreasing spectral density, which, in contrast to many climate indices, usually has the dynamic range of three or more orders of magnitude.

Another important feature of this process is the more noticeable deviation of sampling estimates from the true values. It happens because in this case, the time series containing 100 annual values is very short for getting reliable estimates due

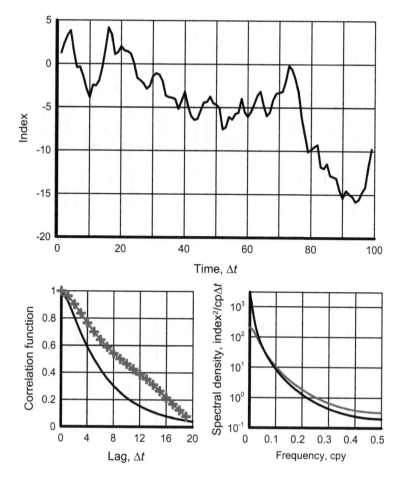

Fig. 2.6 Markov chain of order 2 (common for annual water level of a terminal lake)

to the strong mutual dependence between successive annual values. The number of mutually independent values in this time series is very small thus making the sampling estimates unreliable. This phenomenon plays an important role in determining confidence bounds for estimates of mean values, variances, correlation coefficients, and other statistics (see Chap. 4).

Another example of a second-order Markov chain is the Southern Oscillation Index (SOI). As seen from Fig. 2.7, the correlation function of the simulated SOI time series quickly diminishes to zero, but the spectral density is not monotonic. The ENSO elements (SOI, sea surface temperature in equatorial Pacific, and some ENSO-related climate indices) seem to present rare phenomena in climatic data because their spectra are not monotonic. At the same time, both time series are close to white noise because the spectral peak is rather flat.

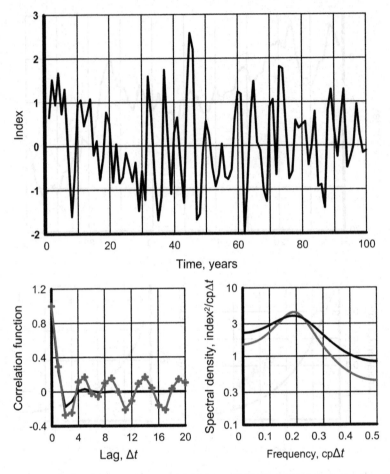

Fig. 2.7 Markov chain of order 2, common for ENSO and ENSO-related phenomena, $\Delta t = 1$ year

In concluding this chapter, it should be noted that even very simple stochastic models of climatic phenomena of the type shown in Figs. 2.5, 2.6 and 2.7 may have complicated spectral densities. For example, the sharply peaked spectrum shown in Fig. 2.8 belongs to a discrete random process that can be completely described with the same simple model of a second-order Markov chain that is described with just two specifically chosen parameters. Climatic processes with strongly peaked spectra seem to be rare.

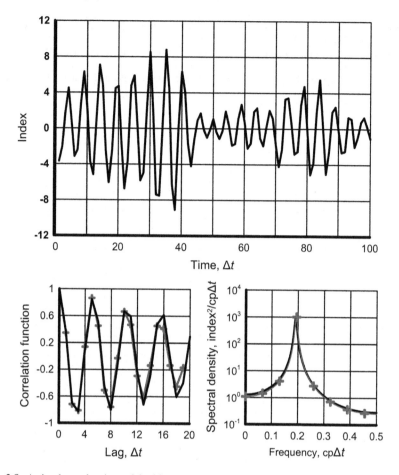

Fig. 2.8 A simple stochastic model with a peaked spectrum

References

Bendat J, Piersol A (2010) Random data. Analysis and measurements procedures, 4th edn. Wiley, Hoboken

Dobrovolski S (2000) Stochastic climate theory. Models and applications. Springer, Berlin

Doob J (1953) Stochastic processes. Chapman & Hall, New York

Duvanin A (1960). Sea tides. Gidrometeoizdat (in Russian)

Hasselmann K (1976) Stochastic climate models, Part I. Theory. Tellus XXVIII:473–485

Monin A (1986) An introduction to the theory of climate. D. Reidel Publishing Company, Dordrecht

Munk W, Cartwright D (1966) Tidal spectroscopy and prediction. Philos Trans R Soc 259(A1105):533–581

Privalsky V (1988) Modeling long term lake variations by physically based stochastic dynamic models. Stochastic Hydrol Hydraul 2:303–315

Wiener N (1948) Cybernetics. John Wiley and Sons Inc., New York

Wiener N (1949) Extrapolation, interpolation, and smoothing of stationary time series, with engineering applications. Wiley, New York
Yaglom A (1962) An introduction to the theory of stationary random functions. Prentice Hall, Englewood Cliffs
Yaglom A (1986) Correlation theory of stationary and related random functions. Basic results. Springer, New York

Chapter 3
Time and Frequency Domain Models of Scalar Time Series

Abstract The most important characteristic of a scalar time series—its spectral density—can be estimated through nonparametric and parametric methods; when the time series is long, the results will be reliable for all mathematically proper methods. The parametric (autoregressive (AR) in this book) approach is relatively reliable with short time series and always provides explicit time domain information in the form of stochastic difference equations, which cannot be obtained through nonparametric methods. The AR approach provides an analytical expression for the spectral density estimate, which is statistically reliable when the AR order is chosen through order selection criteria. Five such criteria are used in this book. The efficiency of autoregressive (or maximum entropy) spectral estimation is demonstrated through parametric and nonparametric analyses of five very short time series whose structures are typical for geophysical data. The AR approach provides an equation for time series extrapolation, for finding its natural frequencies and damping coefficients, and for getting quantitative estimates of time series dependence on its past and upon its innovation sequence. The AR approach is strongly recommended for analysis of geophysical time series. A good nonparametric method of spectral analysis is Thomson's multitaper method.

Generally, it is the covariance function and the spectral density that characterize the behavior of a scalar stationary random process in time and frequency domains. These functions can be calculated directly from the time series by using, in particular, algorithms based upon Eqs. (2.8) and (2.10); the resulting estimates are called nonparametric. Two useful books in this and many other respects are Percival and Walden (1993) and Bendat and Piersol (2010). The statistical properties of estimated correlation function are rather difficult to determine and estimation of correlation functions will not be discussed in this book. The spectral density, or the spectrum, is the function that is necessary for understanding time series properties in the frequency domain and at least partially in the time domain. Therefore, the task of spectrum estimation constitutes an obligatory part of time series analysis.

© Springer Nature Switzerland AG 2021
V. Privalsky, *Time Series Analysis in Climatology and Related Sciences*,
Progress in Geophysics, https://doi.org/10.1007/978-3-030-58055-1_3

3.1   Nonparametric Spectral Analysis

The nonparametric spectral analysis means that a spectral estimate is obtained directly from the time series without making any assumptions about its structure, except for its stationarity or, strictly speaking, its ergodicity. It can be obtained through a Fourier transform of the time series or of its parts with subsequent averaging and smoothing, through a Fourier transform of the covariance function estimate, or by applying a number of filters (windows, or tapers). Such methods are widely used in engineering, where the amount of data is often large and the experiments that generate data can often be repeated at will. The necessary software is easily available in R, MATLAB, and other packages for time series analysis. A thorough review of methods of spectral analysis with practical examples is given in Percival and Walden (1993). The traditional methods of nonparametric spectral analysis used in science and engineering include

- Blackman–Tukey method based upon the Fourier transform of the covariance function estimate plus some tapering (Blackman and Tukey 1958; Bendat and Piersol 1966),
- Bendat–Piersol method of nonoverlapping segments (Bendat and Piersol 2010),
- Welch's overlapped segment averaging method (Welch 1967), and
- Thomson's multitaper method (Thomson 1982).

The first three methods are good for long time series and, with a few exceptions later in this chapter, they will rarely be applied here for analysis of time series, in particular, because so many of them are short. The fourth nonparametric method suggested by David Thomson, can produce, according to the author of the method as well as to this author's experience, unbiased and consistent spectral estimates with short time series; it has high-frequency resolution and can be useful for detecting periodic and quasi-periodic components. In his original publication, the author of the method gave an example of successful analysis of a short ($N = 100$) time series with a rather complicated spectral density typical for sample records in communication systems. A brief explanation for the Thomson's multitaper method (MTM) is that the spectral estimate is obtained as an average of several squared Fourier transforms of the time series which are smoothed with tapers (the so-called discrete prolate spheroidal sequences). This method of spectral analysis will be used in this book along with the autoregressive approach. A review of MTM can be found in Babadi and Brown (2014).

A properly applied direct nonparametric approach provides reliable results when the time series is long, that is, when its length is orders of magnitude longer then the largest time scale of interest. In climatology, the time scales of interest begin with years and do not have an upper limit. The simple rule for the nonparametric methods is that in order to study statistical properties of climate variations with a characteristic time scale of N years, one needs a time series of length $10N$ years or longer. This rule is also correct for any sampling rate and for any other area of research, including all Earth and solar sciences. Thus, the reliability of detecting a spectral peak at 0.04 cpy (a 25-year time scale) from a time series of length 100–150 years using a nonparametric method is dubious especially as the existence of such phenomenon cannot be supported with a physical theory. Therefore, with the exception of the cases when one is interested only in the interannual and higher frequency variability, the above requirement to the length of climate time series is rarely met. Consequently, the nonparametric methods cannot generally produce reliable estimates of correlation functions and spectra in climatology or in any other area where the time series are short. An exception is made here for the Thomson's multitaper method of spectrum estimation.

3.2 Parametric Models of Time Series

The reliability problem with the nonparametric estimation of climate spectra exists due to the necessity to estimate many quantities in order to obtain a detailed and at the same time dependable estimate of the spectrum. When the Blackman–Tukey method is used, one needs to calculate the covariance function at many lags; otherwise, the spectral estimate will have low resolution in the frequency domain. And having many lags means poorer reliability. The Bendat and Piersol and Welch methods require splitting the original time series in as many shorter time series as possible and, at the same time, each subseries should be as long as possible. Therefore, even with a long time series, one has to find a compromise solution between the mutually contradicting desires to get a statistically reliable and, at the same time, high-resolution estimate. Obviously, this difficulty would have been less serious if the number of quantities to be estimated was small in comparison with the number of available observations. This improvement becomes possible with the parametric time series analysis.

The parametric approach arises mostly from the works of Yule (1927) and Wold (1938), who developed the concept of parametric models and introduced the general notion of a random process generated by a linear transformation of a white noise sequence—a linearly regular random process. In the autoregressive model, the current value of the process presents a linear combination of a finite number of its past values plus a "disturbance" consisting of the current value of the white noise sequence. Eventually, it gave rise to several types of parametric models, which are studied in detail in the classical book by Box and Jenkins (1970) and in its four subsequent editions. In this book, only the autoregressive models will be used as the means for obtaining parametric estimates of spectral density.

Though the spectral density function contains some information about the time series behavior in the time domain, it is the parametric approach, which allows one to obtain such information explicitly in the form of stochastic difference equations, with the simplest model being the white noise. In accordance with the definition given above, the time domain model of a stationary Markov chain is described with the following stochastic difference equation of order one:

$$x_t = \varphi_1 x_{t-1} + a_t, \tag{3.1}$$

where the constant $|\varphi_1| < 1$. This equation means that at time t the best prediction of the value x_{t+1} is $\varphi_1 x_t$ because the best estimate of the unknown white noise variable a_t coincides with its mean value, which is supposed to be equal to zero. It can be easily shown that the variance σ_x^2 of the time series (3.1) is

$$\sigma_x^2 = \sigma_a^2/(1 - \varphi_1^2), \tag{3.2}$$

where σ_a^2 is the variance of innovation sequence, which also defines the error variance of forecasting the time series x_t at the unit lead time.

Assuming that the current value x_t depends upon several past values x_{t-k}, $k = 1, \ldots, p$ leads to the stochastic difference equation

$$x_t = \varphi_1 x_{t-1} + \cdots + \varphi_p x_{t-p} + a_t \tag{3.3}$$

or

$$x_t = \sum_{j=1}^{p} \varphi_j x_{t-j} + a_t, \tag{3.4}$$

which is called the autoregressive model of order p, or AR(p).

If the current value x_t depends only upon a linear combination of the current and past values of the innovation sequence a_t and does not depend explicitly upon past values of x_{t-k}, $k > 0$, the respective equation will be

$$x_t = a_t + \sum_{j=1}^{q} \theta_j a_{t-j}. \tag{3.5}$$

This is called a moving average model of order q, or MA(q).

Combining the last two equations leads to the mixed autoregressive and moving average model of order (p, q), or ARMA(p, q):

$$x_t = \sum_{j=1}^{p} \varphi_j x_{t-j} + \sum_{j=0}^{q} \theta_j a_{t-j} \tag{3.6}$$

with $\theta_0 = 1$.

Thus, Eqs. (3.4)–(3.6) define three classes of parametric models of time series: autoregressive AR(p), moving average MA(q), and mixed ARMA(p, q). Their detailed description is available in Box et al. (2015).

The parametric approach to time series analysis is based upon the assumption that the time series can be approximated with an ARMA(p, q) model; an optimal, in some sense, order (p, q) of the model has to be determined and then its parameters—p autoregressive coefficients and q moving average coefficients—are estimated. The variance σ_a^2 of the innovation sequence a_t is found through the estimate of the time series variance and the model's AR and MA coefficients.

The orders p and q of a parametric model can be quite high; in such cases, respective stochastic difference equation will not provide a legible time domain description of the time series. However, if the model's order is not large, say, $p + q \leq 5$, Eqs. (3.4)–(3.6) can give a clear quantitative understanding of how the stationary random process that generated the time series behaves in the time domain, in particular, how strongly the time series depends upon its behavior in the past. These three equations present specific cases of the general regular random process given by Eq. (2.12).

Note also that the autoregressive integrated moving average (ARIMA) models are the same as the ARMA models for the differences $\nabla x_t = x_t - x_{t-1}$ rather than for x_t.

In this book, the only type of parametric models used for time series analysis and forecasting is the autoregressive Eq. (3.4). The equation contains information about two important properties of the random process that it describes. The first property of the AR equation is its ability to determine the degree of predictability of the stationary random process that has generated the time series presented with Eq. (3.4), that is, to describe the degree of success in solving the most important task of time series research: forecasting its behavior. At time t all values of the right-hand part of the equation are known so that the forecasted value at unit lead time for the initial time t depends only upon the innovation sequence term a_t. As the innovation sequence presents a white noise, its forecast for any lead time coincides with the mode of respective PDF which coincides with the mean value in the case of the Gaussian or any other symmetric probability distribution. It means that for any AR(p) model, the prediction error at the unit lead time coincides with a_t and, consequently, the prediction (extrapolation, forecast) error variance at lead time one is σ_a^2. (Prediction of the processes with asymmetrical PDFs is not discussed in this book). When the prediction lead time increases, the prediction error cannot decrease (see Chap. 6), which allows one to introduce a predictability (or persistence) criterion in the form

$$r_e(1) = \sqrt{1 - \sigma_a^2/\sigma_x^2}. \tag{3.7}$$

This criterion defines the degree of success in forecasting the time series at the unit lead time. If the process that has generated the time series is deterministic, the error variance $\sigma_a^2 = 0$ so that $r_e(1) = 1$. If the time series belongs to a white noise process, $\sigma_a^2 = \sigma_x^2$ and $r_e(1) = 0$. For any stationary random process, the criterion $r_e(1)$ satisfies the condition $0 < r_e(1) < 1$. This criterion coincides with the correlation coefficient between the predicted time series value at the unit lead time and the unknown true value.

The other important property of Eq. (3.4) is its ability to determine whether the process contains any oscillations. This is done through the characteristic equation corresponding to the time domain autoregressive model (3.4):

$$1 - \varphi_1 B - \cdots - \varphi_p B^p = 0, \tag{3.8}$$

where B is the backshift operator: $B^j x_t = x_{t-j}$. If the i-th root b_i of this equation is complex-valued, the oscillation e_t can be presented as

$$e_t = e^{-d_i t} \cos(2\pi f_i t - \psi_i), \tag{3.9}$$

where the frequency of oscillation (or the natural frequency of the system)

$$f_i = \frac{1}{2\pi} \tan^{-1}[\mathrm{Im}(b_i)/\mathrm{Re}(b_i)], \tag{3.10}$$

the damping coefficient

$$d_i = |b_i^{-1}|, \tag{3.11}$$

and ψ_i is the phase. If the damping coefficient is large, the oscillation will not be seen in the time series spectrum.

The necessary condition for stationarity of autoregressive process is that the roots b_i of the characteristic Eq. (3.8) should be outside the unit circle.

3.3 Parametric Spectral Analysis

The parametric approach to time series analysis allows one to obtain information about both time and frequency domain properties. By definition, the time series spectrum is the squared modulus of its Fourier transform: $s(f) = |\mathrm{FT}(x_n)|^2$. This way of spectrum estimation is inacceptable because the estimate is not efficient: its variance does not diminish as the time series length increases; however, it can be applied to the time series model, such as AR(p), MA(q), or ARMA(p, q) given with Eqs. (3.4)–(3.6).

To obtain an equation for the spectral density through the autoregressive model, Eq. (3.4) can be rewritten as

$$(1 - \varphi_1 B - \varphi_2 B^2 - \cdots - \varphi_p B^p)x_t = a_t. \tag{3.12}$$

The Fourier transform of this equation is obtained by substituting $e^{-i2\pi j f \Delta t}$ for B^j which leads to the following expression for the spectral density of an autoregressive process of order p:

$$s(f) = \frac{2\sigma_a^2 \Delta t}{\left|1 - \sum_{j=1}^{p} \varphi_j e^{-i2\pi j f \Delta t}\right|^2}, \quad 0 \le f \le f_N. \tag{3.13}$$

where $i = \sqrt{-1}$. This equation means that, up to a multiplier, the spectral density of time series x_t given with an autoregressive model AR(p) of order p is defined with the autoregressive coefficients φ_j, $j = 1, \ldots p$.

Applying the same technique to Eqs. (3.5) and (3.6) leads to the following expressions for spectral densities of the MA(q) and mixed ARMA(p, q) models of time series:

$$s(f) = 2\sigma_a^2 \Delta t \left| 1 - \sum_{j=1}^{q} \theta_j e^{-i2\pi j f \Delta t} \right|^2 \tag{3.14}$$

and

$$s(f) = \frac{2\sigma_a^2 \Delta t \left| 1 - \sum_{j=1}^{q} \theta_j e^{-i2\pi j f \Delta t} \right|^2}{\left| 1 - \sum_{j=1}^{p} \varphi_j e^{-i2\pi j f \Delta t} \right|^2} \tag{3.15}$$

within the frequency range from 0 to f_N. Thus, the shape of the ARMA(p, q) spectrum is completely defined with $p + q$ parameters.

The spectra shown in Chap. 2, with the exception of the white noise model, can be obtained with a good degree of approximation with any of the three models (3.13), (3.14), and (3.15), but describing a moving average or mixed sequence with an autoregressive model requires a high (theoretically, an infinite) AR order p. A similar statement is true for the MA(q) sequence described with AR models.

The properties of the moving average and mixed autoregressive-moving average models are described in many publications, first of all, in Box et al. (2015). Yet, one of the reasons why the autoregressive models are preferable to pure MA or mixed ARMA for describing physical processes in climatic and other geophysical and solar stochastic systems is that they present discrete approximations to the differential equations that are used in fluid dynamics and, in particular, in theory of climate. Besides, the numerical estimation of moving average parameters may present a problem (see Appendix A7.7 in Box et al. 2015).

The autoregressive spectral analysis has another important property. According to the Wiener–Khinchin theorem (2.10)–(2.11), the spectral density presents a Fourier transform of the covariance function. Having a time series of a finite length, one can calculate an estimate of the covariance function $R(k)$ for a finite number of lags $k = 1, \ldots, p$ by using directly the definition of the covariance function given in Chap. 2. The rest of the covariance function at lags higher than p remains unknown. According to Burg (1967), if one requires that the extension of the covariance function beyond the lag p be "the most random," the expression for the estimate of the spectral density will be given with Eq. (3.13). In other words, the AR approach leads to the so-called maximum entropy spectral estimate. ("The most random" is equivalent to "maximum entropy.") According to Jaynes (1982), this estimate possesses a number of useful properties; specifically, it is the smoothest spectrum consistent with the time series, and it would not contain any details that disagree with the available data. This maximum entropy method (MEM) first suggested by J. Burg in 1967 is closely related to the Kolmogorov–Wiener theory of extrapolation of random processes (Chap. 6). A critical review of the method can be found in Percival and Walden (1993). In this book, the autoregressive spectral estimation is regarded as an equivalent to the maximum entropy. This assumption does not play a critical role in the case of time series discussed here because, as shown below, their spectra (and covariance functions) are usually smooth and do not contain sharp and powerful peaks. Yet, whether this approach in spectral analysis is called autoregressive or maximum entropy, the crucial initial step would be to determine the optimal in some sense order p of the model given with Eq. (3.13).

In concluding this section, consider the method often used in climatology and in other sciences under the name of singular spectrum analysis (e.g., Ghil et al. 2002). This method is not easy to categorize by putting it into the class of parametric or nonparametric methods. The reason for this is that the singular spectrum analysis (SSA) is not a method of spectral analysis but rather a way for detecting periodic or quasi-periodic signals of unspecified origin in time series (a signal detection on a background noise). After the time series has been passed through the SSA, it must be subjected to spectral analysis with some traditional parametric or nonparametric method. (Also, see von Storch and Zwiers (1999) where the scalar version of SSA is decrypted as singular system analysis.)

According to the current knowledge, the Earth system does not contain any "signals" in the form of periodic or quasi-periodic components at climatic time scales, with the exception of Quasi-Biennial Oscillation, which is seen with a naked eye and requires no filtering, plus a very weak 19-years long tidal harmonic. Physical considerations and experience show that if a geophysical process has a nonmonotonic spectrum, its shape will be detected by properly conducted spectral analysis. The additional signal detection measures are not necessary. The well-known examples include the Quasi-Biennial Oscillation, ENSO, seasonal and diurnal trends, Madden–Julian Oscillation. The issue of signal detection is hardly productive in climatology and related sciences. These are some of the reasons why the SSA method is not applied in this book.

3.4   Determining the Order of Autoregressive Models

Seemingly, the availability of an exact formula for the spectral density means that the frequency resolution of autoregressive and other parametric spectral estimates is infinitely high because it can be calculated at any frequency. However, as follows from Eq. (3.13), the number of peaks, troughs, and inflection points in autoregressive spectral density estimates cannot exceed the model's order p. Therefore, the AR order is the key parameter that defines the features of an AR (or MEM) spectral estimate. Similar considerations are true for the moving average and mixed ARMA models.

The role of the order p is convenient to characterize with the following simple example. Let $x_t, t = 1, \ldots 100$, be a sample record of a dimensionless white noise process with a unit variance and the sampling interval Δt (e.g., 1 year, 1 day, etc.). According to Eq. (3.13), the true spectrum is a constant: $s(f) = 2\sigma_a^2 \Delta t$. As the true model of the time series is not known at the initial stage of analysis, the spectral estimates should be sought for several values of p, say, from $p = 0$ through $p = 10$ for a time series of length $N = 100$. To select higher values of the AR order would be unreasonable because of common sense considerations: it is not possible to obtain reliable estimates if the number of quantities to be estimated is comparable to the number of observations. The results of analysis of the white noise sequence are shown in Fig. 3.1. Obviously, if one were to choose the AR order $p = 10$ arbitrarily, the conclusion would be that the time series contains "cycles" or quasi-periodic components at frequencies about 0.11 cycles per Δt(cp Δt) and 0.39 cp Δt. However, the correctly defined confidence limits for the estimates given in Fig. 3.1 show that such conclusions would have been false because one can draw a very smooth or even a horizontal line within the confidence interval for the estimate. Obviously, a high order (e.g., $p \geq 10$) does not necessarily mean that the spectral density contains significant peaks. An AR model having a high order may have a very smooth spectrum, while a low-order model can have very sharp peaks (see Fig. 2.8 and the well-known AR(4) example given in the Percival and Walden book published in 1993, pp. 46, 148). Similar examples can be given for more complicated spectra, but the major conclusion is that determining the order of parametric models and showing a confidence interval constitute the absolutely necessary element of parametric spectral analysis.

Several methods can be used to determine the optimal order of an autoregressive model that is being fitted to a time series, but the best approach is to use the order selection criteria (OSC) developed in information theory. Such criteria recommend an optimal order by finding a compromise solution for the following dilemma: a

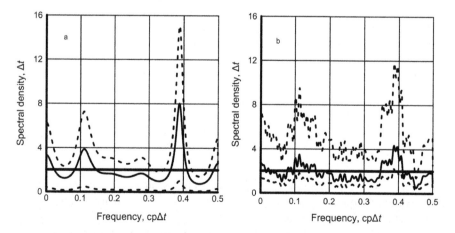

Fig. 3.1 True spectrum of a white noise time series (thick lines) with its AR (or MEM) estimates (**a**) and MTM (**b**) estimates (thick curves) and their 90% confidence limits (dashed lines)

higher order may reveal more details in the spectral density estimate, but at the same time a higher order means that the estimates of AR coefficients are less reliable. The order recommended by order selection criteria is optimal in the sense that it minimizes the variance of innovation sequence—the absolutely unpredictable component of the time series and takes into account the loss of reliability of estimates with the growing AR order. Several such OSCs are known and considered reliable for determining the AR order.

The five criteria used in this book are

- AICc, or corrected Akaike information criterion (Hurwich and Tsai 1989)

$$\text{AICc}(p) \approx \ln[\hat{\sigma}_a^2(p)] + (N + p)/(N - p - 2), \tag{3.16}$$

- RICc, or corrected residual information criterion (Leng 2007)

$$\text{RICc}(p) \approx N \ln[\hat{\sigma}_a^2(p)] + p + \frac{4(p + 1)}{N - p - 2}, \tag{3.17}$$

- BIC, or Schwarz–Rissanen criterion (Schwarz 1978; Rissanen 1978)

$$\text{BIC}(p) \approx \log(\hat{\sigma}_a^2) + p \log(N)/N, \tag{3.18}$$

- PSI, or Hannan–Quinn criterion (Hannan and Quinn 1979)

$$\text{HQC}(p) \approx \log(\hat{\sigma}_a^2) + 2p \log(\log(N))/N, \tag{3.19}$$

- CAT, or Parzen criterion of autoregressive transfer function (Parzen 1974)

$$\text{CAT}(p) \approx \frac{1}{N} \sum_{j=1}^{p} \hat{\sigma}_a^{-2}(j) - \hat{\sigma}_a^{-2}(p). \tag{3.20}$$

Here, $\hat{\sigma}_a^2$ is the estimated variance of the innovation sequence a_t. The approximate equality turns to equality if the time series is generated by a Gaussian random process. The criteria should be calculated for AR orders from 0 (when $\sigma_a^2 = \sigma_x^2$) to some maximum order $p_{max} \ll N$. The optimal order corresponds to the minimal value of the criterion. Other versions of order selection criteria are possible, but all of them

Table 3.1 Efficiency of order selection criteria

True order	N	AICc	CAT	BIC	PSI
0	100	0.69/0.80	0.70/0.81	0.96/0.99	0.88/0.95
0	1000	0.72/0.83	0.72/0.83	0.99/1	0.94/0.98
9	100	0.37/0.62	0.36/0.60	0.03/0.06	0.18/0.30
9	1000	0.71/0.82	0.79/0.82	0.99/1	0.94/0.98

The numbers after the slash show results including AR orders $p \pm 1$

usually produce satisfactory results. Examples can be found in Shi and Tsai (2001), in Shumway and Stoffer (2017), and in MATLAB.

Results of a series of Monte Carlo experiments with AR models having orders from 0 through 5 (plus a case when the order was equal to 24) were given in Privalsky and Yushkov (2018). That work has been dealing with climate data that is, with a situation when higher orders and sharply peaked spectra are rare. Some additional information intended for general situations in geophysics are given Table 3.1.

Actually, the proper choice of model's order is always necessary but not always critical. The absolute requirement here is not to select a high order arbitrarily, without a proper testing. This is what the order selection criteria are for. They are quite reliable for selecting a proper order, and an error by ± 1 is not necessarily critical if the selected order is higher than $p = 2$. Then, an error by one will not lead to critically erroneous results. The criteria rarely indicate orders much higher than the true one.

If the true order is zero (white noise), the probability that any of the criteria will select $p = 10$ or higher is low. Most choices will be zero or one. This is what can be seen in Table 3.1. When the process is close to white noise, all spectral estimates will show this property even in those few cases when the order is exaggerated.

If the unknown true AR order of a geophysical time series is high, the situation may change. For example, if $p = 9$ such as for the annual sunspot numbers (Chap. 13), the criteria will work well for longer time series ($N = 1000$) but will fail if the time series is short ($N = 100$). The AR order will be wrong in many if not practically all cases (Table 3.1). In the scalar case, the most important statistics of the time series is its spectrum, and one should find out if the erroneously selected low order leads to an erroneous spectral estimate. If the process is cyclic (Fig. 3.2a), even the wrong estimate of the spectrum may not be catastrophic. The true order of the time series is $p = 9$, but the erroneously selected order $p = 2$ still results in a spectral estimate that shows the generally correct type of the spectral density (Fig. 3.2b). Thus, in cases when the time series behaves in a specific manner such as showing a cyclicity seen with a naked eye, even a serious error in the AR order recommended for the time series may not be catastrophic for estimation of its spectral density. Generally, if the model is complicated, that is, if its order is high and the spectrum contains a sharp peak, a moderate distortion of the spectral estimate is a payment for the lack of sufficient data.

In concluding this section, consider another useful feature of autoregressive models. The AR equation can be used to determine contributions of time series'

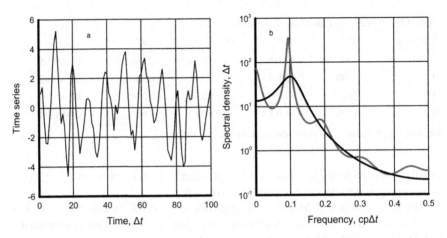

Fig. 3.2 A cyclic time series of an AR(9) model (**a**), and its spectra (**b**): the true spectrum (gray line) and the estimate obtained from a time series of length $N = 100$ and AR order $p = 2$ (black line)

past values to its variance σ_x^2. Multiplying Eq. (3.3) by x_t and finding the mathematical expectation of the products, one gets the equation for the time series variance as a function of its covariance function, AR coefficients, and the innovation sequence variance:

$$\sigma_x^2 = \varphi_1 R(k) + \cdots + \varphi_p R(p) + \sigma_a^2. \qquad (3.21)$$

The summands in the right-hand part of this equation present contributions of the p past values of x_t to the variance σ_x^2. This information can be useful for understanding the time domain behavior of the time series.

3.5 Comparison of Autoregressive and Nonparametric Spectral Estimates

The statement that the spectral density is the most important statistics of any scalar time series is true for Gaussian and non-Gaussian data. The spectral density can be estimated with parametric and nonparametric methods, and if the time series is long, the estimates will be reliable and similar to each other. If the time series is short, which happens regularly in climatology and its branches and in other Earth sciences, the task of proper spectral analysis becomes vital. In this section, we will show the advantages of the parametric approach to the task of spectral estimation, which exist, in particular, due to the fact that many geophysical processes can be well approximated with stochastic difference equations of relatively low order. The examples given here include time series whose spectra are typical for climatic processes, including the annual global surface temperature. The time series will always be short: the total number of its terms is just 50. The autoregressive spectral estimates will be compared with respective nonparametric estimates obtained according to Blackman and Tukey (1958). Four time series of length $N = 50$ have AR orders from 1 to 4, and they are quite common for climate and for other geophysical phenomena. The true AR models are known in all cases, and the initial data for the analysis were obtained through simulation. Because of the small length of the time series, the sample estimates of AR coefficients may differ rather significantly from the true values.

The true model of the first time series is an AR(1) with the AR coefficient $\varphi = 0.5$. It can be the annual river streamflow, daily temperature, sea level variations in coastal areas with no tides, etc. The sample AR (or MEM) estimate of the spectrum is shown in Fig. 3.3a along with the true spectrum.

According to Fig. 3.3a, the 90% confidence interval for the AR spectral estimate contains the true spectrum; this estimate can be regarded as satisfactory. The shape of the spectrum is reproduced accurately, and the bias occurs due to the sampling variability of the variance estimate. The nonparametric estimate (Fig. 3.3b) has several peaks, but the peaks are statistically insignificant. The 90% confidence interval is wide and asymmetric with respect to the spectral estimate due to the asymmetry of χ^2 distribution at low degrees of freedom in this and the other three examples below.

More complicated AR models are shown in Figs. 3.4, 3.5, and 3.6. In the first case, the true model is AR(2) with AR coefficients $\varphi_1 = 0.5$ and $\varphi_2 = -0.4$. The true spectrum is close to the spectrum of the Southern Oscillation Index. The characteristic equation of this model is

$$1 - 0.5B + 0.4B^2 = 0 \qquad (3.22)$$

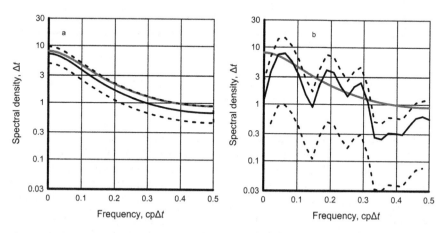

Fig. 3.3 Autoregressive (**a**) and Blackman and Tukey (**b**) spectral density estimates of the AR(1) time series, $N = 50$. The gray line shows the true spectrum

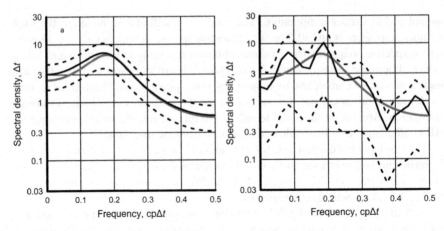

Fig. 3.4 Autoregressive (**a**) and Blackman and Tukey (**b**) spectral density estimates of an AR(2) time series, $N = 50$. The gray line shows the true spectrum

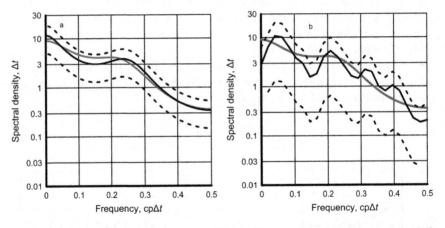

Fig. 3.5 Autoregressive (**a**) and Blackman and Tukey (**b**) spectral density estimates of an AR(3) time series, $N = 50$. The gray line shows the true spectrum

and it has two complexly conjugated roots corresponding to the natural frequency $f_e \approx 0.19 \, \Delta t^{-1}$ and a damping coefficient $d \approx 0.63$. Again, the AR (or MEM) spectral estimate (Fig. 3.4a) is definitely better than the nonparametric one (Fig. 3.4b).

In this case, only two parameters are to be estimated. With $N = 50$, the estimates are relatively reliable and the sample spectral estimate is quite close to the true one (Fig. 3.4a). The nonparametric estimate hints that the spectrum is nonmonotonic and possibly contains a smooth maximum, but actually the confidence band is so wide that one can draw a practically monotonic line inside it.

The true values of AR coefficients for the more complicated AR(3) and AR(4) models are [0.73 −0.41 0.21] and [0.57 −0.02 0.07 0.23], respectively. In both

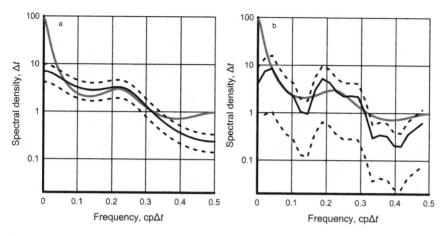

Fig. 3.6 Autoregressive (**a**) and nonparametric (**b**) spectral density estimates of an AR(4) time series, $N = 50$. The gray line shows the true spectrum

cases, the natural frequency of the system is approximately $0.23\, \Delta t^{-1}$. The estimate for the AR(3) time series, which simulates the Palmer Draught Severity Index over the contiguous USA, stays close to the true spectral density for the AR(3) time series (Fig. 3.5a) and becomes less accurate for the AR(4) time series (Fig. 3.6a). However, the shape of the spectrum is reproduced rather accurately including the hump at intermediate frequencies. This simulated time series is similar to the actual annual global surface temperature (Privalsky and Yushkov 2018). In all cases, the nonparametric estimates given in Figs. 3.5b and 3.6b are less accurate, but one has to remember that the time series are very short for the nonparametric spectral estimation to be efficient. The incorrect behavior of the spectral estimates at low frequencies in Fig. 3.6 is the results of a poor estimate of the time series variance due to its short length.

These examples are given here to illustrate the following argument: when the optimal autoregressive model of a geophysical or any other time series indicated by order selection criteria is low, the autoregressive approach allows one to obtain statistically reliable estimates of AR coefficients and, consequently, of the spectral density even when the time series is very short. Also, a low AR order does not necessarily mean that the spectrum of the time series does not contain any sharp peaks.

Comparing Figs. 3.3a through 3.6a with Figs. 3.7 and 3.8 reveals the high quality of Thomson's MTM spectral estimates even with very short time series; however, MTM does not provide any explicit time domain information about the time series.

The above given examples describe relatively simple AR models whose spectral densities do not contain significant peaks. Consider now an AR(18) model with a high dynamic range of three orders of magnitude. Its spectrum is shown in Fig. 3.9 up to the frequency $0.1\, \Delta t^{-1}$; it contains one rather smooth peak at $0.022\, \Delta t^{-1}$ and then monotonically decreases at higher frequencies. If $\Delta t = 1$ day, this model simulates

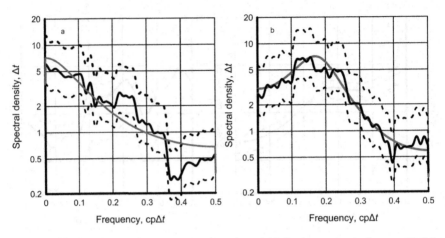

Fig. 3.7 MTM spectral density estimates of AR(1) and AR(2) time series (**a**, **b**, respectively), $N = 50$. The gray line shows the true spectrum

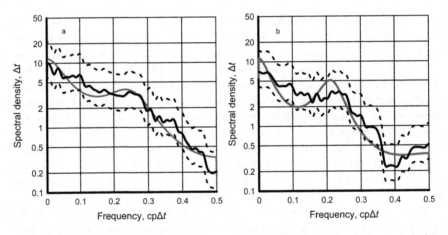

Fig. 3.8 MTM spectral density estimates of AR(3) and AR(4) time series (**a**, **b**, respectively), $N = 50$. The gray line shows the true spectrum

the Madden–Julian Oscillation. According to the rule of selecting the maximum order not higher than approximately one-tenth of the time series length, the length of the simulated time series was put to 200.

The AR estimate of the spectral density obtained for this time series is $p = 6$, which is quite different from the true order $p = 18$. Yet, the shape of the estimated spectral density is correct though the frequency of the peak is determined with a 15% bias (Fig. 3.9a). The MTM spectral estimate shown in Fig. 3.9b is also rather approximate. This example shows that when the true model is complicated and the data are scant, the quality of the AR (or MEM) spectral estimation may not be very high. This is just a common sense consideration.

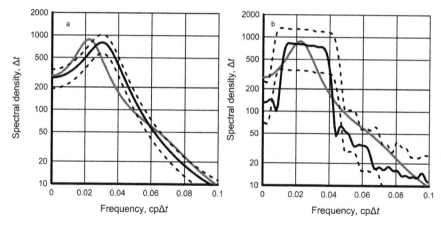

Fig. 3.9 AR (MEM) and MTM spectral estimates of the AR(18) spectrum (**a**, **b**, respectively), $N = 200$. The true spectrum is shown with the gray line

3.6 Advantages and Disadvantages of Autoregressive Analysis (Scalar Case)

The autoregressive approach provides quantitative information about statistical properties of time series in both time and frequency domains. This can also be achieved with other parametric models, but the moving average operator is less reasonable physically and is difficult to deal with. The nonparametric methods of analysis do not produce explicit information about the time series behavior in the time domain and, if the time series is short, the nonparametric spectral estimates are less reliable than the AR estimates.

The autoregressive stochastic difference equation possesses a number of useful features for analysis of stationary time series:

- it presents a ready-to-use tool for linear extrapolation of time series (Chap. 6);
- it allows one to get quantitative estimates of time series dependence upon its previous values (up to the AR order p) and upon the innovation sequence;
- it can be used to determine the natural frequencies and damping coefficients of the time series, that is, the frequencies, at which the spectral density may have peaks;
- it provides an analytical expression for the time series spectral density;
- the autoregressive approach to spectral estimation satisfies the requirements of the maximum entropy method (MEM), and it is capable of producing satisfactory estimates of the spectrum when the time series is short.

This author is not aware of any disadvantages in applying the autoregressive time and frequency domain approach to time series research in climatology and other Earth and solar sciences.

References

Babadi B, Brown E (2014) A review of multitaper spectral analysis. IEEE T Bio-Med Eng 61:1555–1564

Bendat J, Piersol A (1966) Measurement and analysis of random data. Wiley, New York

Bendat J, Piersol A (2010) Random data. Analysis and measurements procedures, 4th edn. Wiley, Hoboken

Blackman R, Tukey J (1958) The measurement of power spectra from the point of view of communication engineering. Dover Publications, New York

Box G, Jenkins G (1970) Time series analysis. Forecasting and control. Wiley, Hoboken

Box G, Jenkins M, Reinsel G, Liung G (2015) Time series analysis. Forecasting and control, 5th edn. Wiley, Hoboken

Burg JP (1967) Maximum entropy spectral analysis. Paper presented at the 37th Meeting of Society for Exploration Geophysics. Oklahoma City, OK, October 31, 1967

Ghil M, Allen M et al (2002) Advanced spectral methods for climate time series. Rev Geophys 40:1–41

Hannan E, Quinn B (1979) The determination of the order of an autoregression. J R Stat Soc B 41:190–195

Hurwich M, Tsai C-L (1989) Regression and time series model selection in small samples. Biometrica 76:297–307

Jaynes E (1982) On the rationale of the maximum entropy method. Proc IEEE 70:939–952

Leng C (2007) The residual information criterion, corrected. http://arxiv.org/abs/0711.1918

Parzen E (1974) Some recent advances in time series modeling. IEEE T Automat Contr 19:723–730

Percival D, Walden A (1993) Spectral analysis for physical applications. Cambridge University Press, Cambridge

Privalsky V, Yushkov V (2018) Getting it right matters: climate spectra and their estimation. Pure Appl Geoph 175:3085–3096

Rissanen J (1978) Modeling by shortest data description. Automatica 14:465–71

Schwarz G (1978) Estimating the dimension of a model. Ann Stat 6:461–464

Shi P, Tsai C-L (2001) Regression model selection—a residual likelihood approach. J Roy Statist Soc B 64:237–252

Shumway R, Stoffer D (2017) Time series analysis and its applications, 4th edn. Springer, Heidelberg

Thomson D (1982) Spectrum estimation and harmonic analysis, P. IEEE 70:1055–1096

von Storch H, Zwiers F (1999) Statistical analysis in climate research, 2nd edn. Cambridge University Press, Cambridge

Welch P (1967) The use of Fourier transform for the estimation of power spectra: a method based on time averaging of short, modified periodograms. IEEE T Acoust Speech 15:70–73

Wold H (1938) A study in the analysis of stationary time series. Almqvist and Wiksells BokTryckeri-A-B, Uppsala

Yule G (1927) On a method of investigating periodicities in disturbed series, with special attention to Wolfer's sun spots. Philos Trans R Soc Lond A CCXXVI:267–298

Chapter 4
Practical Analysis of Time Series

Abstract After a preliminary processing, the time series should be tested for stationarity. The test may fail if the time series contains a trend or if its mean value, variance, or spectrum are found to be time-dependent. Deleting the trend can be justified if it is caused by external factors or if it interferes with the higher-frequency part of the spectral density of interest to the researcher. A test for stationarity is suggested through splitting the time series in halves and estimating the mean values, variances and, if possible, spectral densities of the entire time series and its halves. The confidence bounds for estimates of statistical moments depend upon the number of independent observations in the time series. These numbers depend upon the correlation structure of the time series, and they can be much smaller than the total number of terms in the time series. A linear filtering is generally not recommended. The autoregressive approach allows one to determine frequencies of even strictly periodic oscillations contained in the time series (tides) with exceptionally high accuracy providing that the time series is long. A detailed example of autoregressive analysis is given.

The two mandatory requirements in practical analysis of time series are

- using the proper methods of analysis and
- calculating confidence intervals for all estimated statistical characteristics.

   This means, in particular, that the methods of time and frequency domain analysis should be mathematically suitable for the time series which is being analyzed. In this book, the fundamental approach to analysis is based upon autoregressive modeling, which has the ability to provide relatively reliable estimates of time series statistics even with short time series. Its other advantage is the explicit time domain model which is obtained at the initial stage of autoregressive analysis and which does not exist if a nonparametric approach is used. It will be shown later in this chapter (Sect. 4.5) that the autoregressive approach can be quite effective for estimating the spectral density of time series with a very complicated structure.

   If the time series is short, its spectrum should not be estimated with nonparametric methods other than Thomson's MTM. Moreover, all statistical estimates should be accompanied with respective reliability estimates. Any estimate of statistical characteristics is absolutely useless if it is not supplemented with confidence intervals or

© Springer Nature Switzerland AG 2021
V. Privalsky, *Time Series Analysis in Climatology and Related Sciences*,
Progress in Geophysics, https://doi.org/10.1007/978-3-030-58055-1_4

some other quantitative indicator of its reliability in accordance with mathematical statistics. If confidence intervals or reliability estimates are available but not shown intentionally, their absence should be explained.

If the time series contains outliers, they should be handled before continuing the analysis. The next step is to build an autoregressive model of the time series, including respective spectral estimate, and to check for any unexpected features such as a significant low-frequency trend and/or unexplainable spectral peaks in the model selected by most order selection criteria. The latter problem is rare because normally the order selection criteria do not allow unreasonably high or low orders. The information obtained at that stage should be used to decide if any additional steps might be required.

The common problems of this type include

- an incorrect sampling interval,
- the presence of a linear or nonlinear trend in the time series,
- the presence of strong high-frequency and/or quasi-periodic fluctuations in the AR model' spectrum, which need to be explained.

The trend can be a product of natural low-frequency components or it can be caused by some artificial external forcing (such as an anthropogenic effect). The decision whether to delete the trend or analyze the time series as is should not be taken without a justification.

4.1 Selecting the Sampling Interval

The sampling interval Δt should be set in agreement with the task of the analysis. If it is the climate variability at time scales longer than 2.5–3 years, the interval $\Delta t = 1$ year is generally sufficient. The resulting highest frequency in the time series spectrum is the Nyquist frequency $f_N = 1/2\Delta t$, that is, 0.5 cpy when $\Delta t = 1$ year. The frequencies that can be analyzed reliably begin from approximately 0.30–0.35 cpy. Simple interpolation between consecutive terms of the time series intended to get a faster sampling rate is useless. If the spectrum is expected to contain high energy at higher frequencies, the interval should be smaller; such cases are rather rare in Earth sciences. Setting $\Delta t = 1$ month for studying climate variability is normally not reasonable, in particular, because it may transform a stationary time series into a sample of a periodically correlated (cyclostationary) random process. Also, an exceedingly small sampling interval creates redundant information and reduces the spectral resolution at lower frequencies. The general rule here is that the time series should contain at least several measurements per the smallest time scale of interest. For example, the choice of $\Delta t = 1$ year could be too large for studying the Quasi-Biennial Oscillation whose characteristic time scale is approximately 2.3 year. These considerations are relevant for other conditions when the time series is not related to climate and when time is measured in seconds, hours, or any other units.

Detailed recommendations for setting the Nyquist frequency when dealing with recording and/or preparing time series for further analysis are given in Chap. 10 of the Bendat and Piersol book (2010) and in the book by Thomson and Emery (2014).

4.2 Linear Trend and Its Analysis

The trend in the time series can be linear or nonlinear, and it is reasonable to assume that its presence will affect results of analysis at higher frequencies, that is, at smaller time scales. The presence of a trend raises at least two questions: what is the cause of the trend and should it be removed? One also needs to understand how the removal of the trend will affect statistical properties of the time series. The considerations given below can be valid for nonlinear trends as well but only the linear case will be discussed here.

Figure 4.1 shows anomalies of the annual global terrestrial temperature from 1856 through 2017 (see #1 in Appendix to this chapter). Obviously, there is a substantial trend in the curve, which makes the spectral density high at low frequencies (Fig. 4.2). If the trend is deleted, the spectral density diminishes by an order of magnitude at low

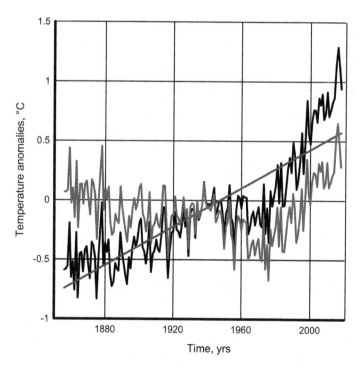

Fig. 4.1 Anomalies of annual land surface temperature before (black) and after (gray) linear trend deletion, 1856–2017

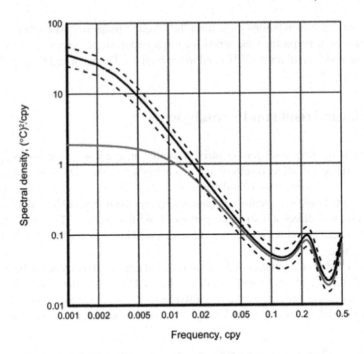

Fig. 4.2 Spectra of land temperature anomalies before (black) and after (gray) linear trend deletion

frequencies, but the two estimates of the spectrum stay quite close to each other at frequencies higher than 0.02–0.03 cpy, that is, at time scales less than approximately 50 years. Obviously, a time series of length $N \Delta t = 162$ years, as in this case, is too short to obtain reliable information about the behavior of temperature at such time scales so that trend deletion would not affect the spectrum of temperature at relevant frequencies.

On the other hand, trend deletion affects statistical predictability of the time series: if the trend is caused by natural climate variability, it should stay as a part of the time series; otherwise, the predictability will be reduced artificially.

The next step would be to decide whether the time series can be treated as a sample of a stationary or even ergodic, random process.

4.3   Testing for Stationarity and Ergodicity

The only assumption, which is made about time series for its standard statistical analysis, including spectral estimation, is that it belongs to a stationary random process. To prove with probability one that the assumption is correct is not possible but one can verify whether it can be acceptable for a specific time series. This may be done in the following manner:

- split the time series in two equal parts and
- verify that the differences between the statistics of the entire time series and its halves do not lie outside the limits of sampling variability of respective estimates for the entire time series.

If the results of such verification are favorable, that is, if the differences can be ascribed to the sampling variability, there seems to be no ground to reject the initial assumption of stationarity. If, in addition, the probability distribution of the time series is Gaussian, the hypothesis of ergodicity that is usually accepted for stationary time series by default becomes reasonable as well. Comparisons should include at least the first two statistical moments, that is, the mean value and variance (or the root mean square value). The variances of estimated mean values and estimated variances should be calculated with account for the number of statistically independent observations in the time series.

For the time series $x_t, t = 1, \dots, N$, the root mean square (RMS) error of the estimated mean value $\bar{x}$ is

$$\sigma[\bar{x}] \approx \sigma_x / \sqrt{\tilde{N}} \tag{4.1}$$

where σ_x is the estimated standard deviation and $\tilde{N}$ is the effective number of independent observations in the time series:

$$\tilde{N} = N / \sum_{k=-\infty}^{k=\infty} r(k) \tag{4.2}$$

where $r(k)$ is the correlation function of the time series.

The RMS error of the estimated variance σ_x^2 is

$$\sigma[\sigma_x^2] = \sigma_x^2 / \hat{N}, \tag{4.3}$$

where

$$\hat{N} = N / \sum_{k=-\infty}^{\infty} r^2(k). \tag{4.4}$$

For more details, see Yaglom (1987). The formulae with the total number of observations N used instead of $\tilde{N}$ and $\hat{N}$ are correct only if x_n is a white noise sample. Generally, such an assumption is wrong and is not applicable to time series.

Thus, the test for stationarity with respect to the mean value and variance includes the following steps:

- fit a proper AR model to the entire series and to its halves.
- estimate mean values and variances for each of the three time series.

If the confidence limits for the estimates of mean values overlap and the same is true with respect to the variance (or RMS) estimates, the hypothesis of stationarity should not be rejected.

Using Eqs. (4.2) and (4.4) requires the knowledge of the correlation function $r(k)$. It can be obtained for any maximum lag in the following manner. When the time series is approximated with an AR(p) model, the first p values of $r(k)$ coincide with the sampling estimates obtained directly through the standard formulae (2.8) and (2.9). The extension of the correlation function of an AR model of order p is calculated as

$$r(k) = \varphi_1 r(k-1) + \cdots + \varphi_p r(k-p), k = p+1, \ldots, K, \qquad (4.5)$$

where the number K is large enough so that the absolute values of $r(k)$ become close to zero when $k > K$. The sequence given with Eq. (4.5) presents the maximum entropy extension of correlation function estimated for values of $k \leq p$ through Eqs. (2.8) and (2.9). The recommended value of K for time series whose spectrum is not concentrated in a narrow frequency band is $K = 100$. Otherwise, it should be increased. For the example given below, $K = 500$.

Strictly speaking, the spectral density estimates should also be included into the procedure of testing for stationarity, but if the time series is short or the spectral density estimate is not complicated, the spectrum test can be omitted.

Consider an example using the time series of monthly sunspot numbers (SSN) from 1749 through 2018 (see #2 in Appendix and Fig. 4.3).

The total number of observations is $N = 3240$ for the entire data set and $N = 1620$ for its halves SSN1 and SSN2. The optimal AR models chosen by OSCs for the time series are shown in Table 4.1. According to the table, the numbers of independent observations $\tilde{N}$ and $\hat{N}$ are more than an order of magnitude smaller than the number of observations N. It happens because the time series has a cyclic nature and the integrals of its correlation function and correlation function squares are large. This is just one example showing the importance of determining the number of mutually independent (or mutually uncorrelated) observations in the time series that are more complicated than the white noise process.

It follows from the table that the three 90% confidence intervals for estimates of mean value and standard deviation overlap so that the three sample estimates for the mean value do not differ statistically significantly from each other. The same statement is true for the estimates of the time series root mean square value. Thus, the assumption of stationarity of the SSN time series does not contradict the data.

A stricter test for this time series that has a complicated spectrum would be a comparison of the three spectral estimates obtained from the entire time series SSN and from its halves SSN1 and SSN2 (Fig. 4.4). As seen from the figure, the three spectral estimates are similar and to so that the assumption of stationarity is acceptable from this point as well. (The confidence intervals in this case are not required.)

The SSN variations are limited by zero from below and behave like a quasi-periodic process. Their PDF cannot be Gaussian so that the ergodicity of the random process that generated the SSN time series cannot be verified.

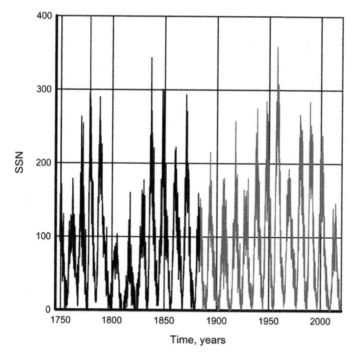

Fig. 4.3 SSN time series split into halves

Table 4.1 Stationarity test for the SSN time series, 1749–2018

Time series	N	p	Mean value		RMS	
			$\tilde{N}$	Confidence interval	$\hat{N}$	Confidence interval
SSN	3240	34	137	[72.8 82.4 91.9]	109	[61.1 67.8 76.3]
SSN1	1620	25	51	[65.2 80.8 96.4]	61	[58.0 66.5 78.2]
SSN2	1620	29	67	[69.9 84.9 98.0]	53	[55.2 69.1 82.7]

It should be stressed that the above described test presents an approximate technique, and it cannot be regarded as an absolute proof of stationarity.

4.4 Linear Filtering

Time series are often subjected to filtering designed isolate variations within specific frequency bands. Generally, the spectra of climate and many other geophysical data are smooth, and the only frequency band that dominates the spectrum is at the lower end of the frequency axis. This means that if one wants to "protect" some specific band within the spectrum from variations belonging to a different band, filtering

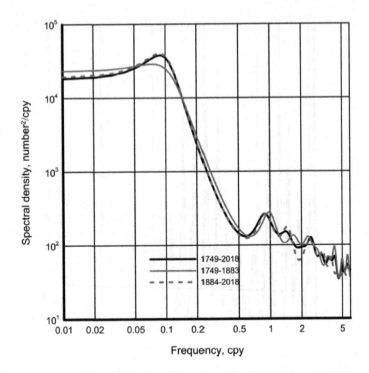

Fig. 4.4 Spectral estimates for the SSN time series

is not required. Moreover, with autoregressive (maximum entropy) spectral estimation, there is no interaction between different frequencies, which makes the filtering operation unnecessary or even harmful.

If the goal of filtering is to study variations within a specific frequency band in the time domain, one should remember that statistical properties of the initial and filtered time series are very different. In particular, if a filter which suppresses high-frequency components is applied, the numbers of mutually uncorrelated observations will be smaller and the reliability of all estimates will be worse than it was before the filter has been applied to the time series. The filtered time series has to be analyzed to determine variances of estimates obtained from it. In short, time series should not be filtered unless one has strong physical and/or probabilistic arguments in support of the filtering operation.

There are three types of filters:

- the low-pass filter removes high-frequency (fast) fluctuations.
- the high-pass filter removes low-frequency (slow) fluctuations.
- the band-pass filter removes fastest of the slow fluctuations and slowest of the fast fluctuations.

When passing a time series through a filter, one should have in mind the following factors:

- no physically realizable filter can remove fluctuations in a given frequency band without affecting all other frequencies.
- a longer weighting function of the filter means a narrower frequency band to which the filter is tuned.

The filtering operation is done in the time domain in accordance with the formula

$$\tilde{x}_t = \sum_{k=-K}^{K} \lambda_k x_{t+k}, \tag{4.6}$$

where λ_k is the weighting function of the filter. It makes the time series (4.6) shorter than the initial time series by $2K$ terms. This latter effect can be avoided by building an AR model of the time series prior to its filtering and then simulating it at both ends of the time series for sufficiently long intervals. Then, the filter is applied to the resulting longer time series, which can now be studied within the entire time interval for which the observation data were available initially.

The properties of the filter in the frequency domain are defined by the filter's frequency response function (FRF) which presents a Fourier transform of the weighting function:

$$H(f) = \sum_{k=-K}^{K} \lambda_k e^{-i2\pi k f \Delta t}, \tag{4.7}$$

where $i = \sqrt{-1}$. The spectral density of the time series that passed through a filter is transformed from $s(f)$ to

$$\tilde{s}(f) = |H(f)|^2 s(f). \tag{4.8}$$

This means, in particular, that if the time series x_t is generated by a white noise process whose spectrum is constant over the entire frequency interval from 0 to $1/2\Delta t$, the spectrum of the time series at the output of the filter coincides with the square of filter's frequency response function. For example, the FRF of the low-pass filter with an equal-weight function λ_k, which is often used in climatology and geophysics in general, has the form

$$H(f) = \frac{\sin \pi f K \Delta t}{\pi f K \Delta t}, \quad 0 \le f \le 1/2\Delta t. \tag{4.9}$$

Examples of this FRF are given in Fig. 4.5.

The low-pass filtering (smoothing) is convenient for showing the slower variations of the time series which can be masked in the original time series with contribution from high frequencies. An example is given in Fig. 4.6 which shows the anomalies of annual surface temperature over the World Ocean from 1850 through 2017 along with the same time series at the output of a low-pass filter with the half-length of

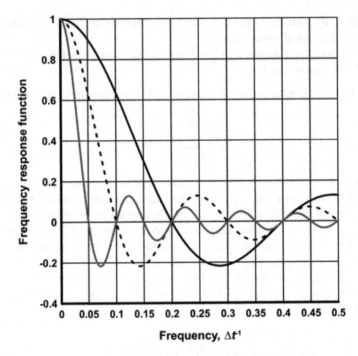

Fig. 4.5 Frequency response function (4.9) for $K = 5$ (black), $K = 10$ (dashed), and $K = 20$ (gray)

the weighting function equal to 10 years (see #1 in Appendix). This operation has resulted in the loss of the first and last ten terms of the original time series, but the low-frequency behavior of the time series became clearer.

The frequency response function of a high-pass filter can be presented as $\hat{H}(f) = 1 - H(f)$, where $H(f)$ is the FRF of the low-pass filter. Therefore, the high-pass filtered time series $\hat{x}_t$ can be obtained as the difference between the original and the low-pass filtered time series:

$$\hat{x}_t = x_t - \tilde{x}_t, t = K + 1, \ldots, N - K \tag{4.10}$$

Similarly, high-frequency fluctuations or fluctuations within an intermediate frequency band can be isolated with high- and band-pass filters. Yet, one must always remember that the filtering operation affects the entire spectrum of fluctuations and can never be ideal. Generally, time series filtering is not required and not recommended for analyzing geophysical time series. In any case, the necessity to apply a filter to the original time series must be proved.

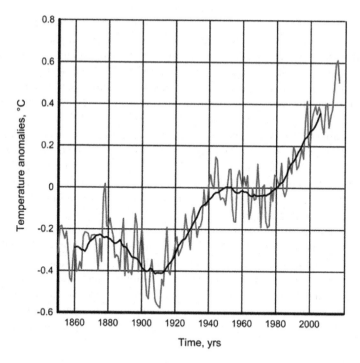

Fig. 4.6 Observed (gray) and smoothed (black) variations of annual surface temperature anomalies, World Ocean, 1850–2017

4.5 Frequency Resolution of Autoregressive Spectral Analysis

The AR (or MEM) spectral estimation provides an analytical formula for the estimated spectrum. It means that the spectral resolution in the formula is such that the value of spectral density can be calculated at any frequency. This is true, but the actual resolution is defined by the AR order: the number of extrema and inflection points in the spectral curve corresponding to an AR(p) model cannot be higher than p (see Sect. 4.3). Therefore, a high resolution requires a high AR order, but a high-order model cannot be obtained with a short time series.

By definition, a linearly regular random process does not contain any strictly periodic components. This feature may cause some doubts about the ability of parametric time series analysis designed for regular processes to detect sharp peaks at frequencies which are close to each other, for example, when the data contain harmonic oscillations. Actually, the ability of autoregressive spectral analysis in this respect is very high under just one condition: getting accurate results requires having enough data for analysis. (Certainly, this requirement holds for all nonparametric method of spectral analysis such as Blackman and Tukey's, MTM, Welch's, etc.)

A unique case of harmonic oscillations with perfectly known frequencies within the Earth system is tides. The frequencies of tidal constituents are known precisely from astronomy; the amplitudes are determined from observations. The autoregressive analysis in the frequency domain provides a convenient tool for estimating frequencies of harmonic oscillations that are contained in time series of tidal phenomena. If the frequencies are determined correctly in sea level observations, one may hope that they will also be determined correctly in any other stationary data.

The example below is designed to verify how accurately the maximum entropy spectral analysis can determine the frequencies of tidal constituents by analyzing the time series of sea level at station 9414317, Pier 22 ½, San Francisco, USA, using 10^5 hourly sea level observations starting from January 28, 2000. The data source is #3 in Appendix. A part of the record (about 50 days) is shown in Fig. 4.7. The tides obviously dominate the record.

The frequencies of the main tidal constituents were determined by conducting autoregressive spectral analysis of the entire time series of length 10^5 h, the AR order $p = 10^4$, and the frequency resolution of the spectral estimate 10^{-6} cph. The results for the diurnal tides are shown in Fig. 4.8 and in Table 4.2.

As seen from the table, the errors in estimates of the constituents' periods do not exceed 0.022%. The average error for the entire range of diurnal tides is 0.0077%. This is not a misprint; it is a proof that the autoregressive spectral analysis does

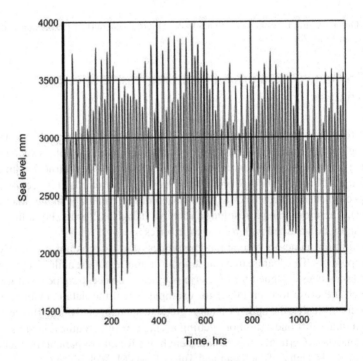

Fig. 4.7 Sea level variations at Fort Point, San Francisco, USA

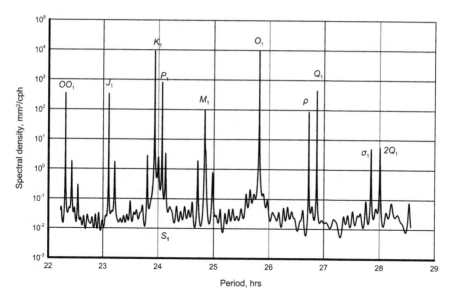

Fig. 4.8 Autoregressive spectral estimate of sea level variations at Fort Point, San Francisco, USA, within the frequency band of diurnal tides

Table 4.2 Diurnal tidal constituents at sea level station 9414317, San Francisco, USA (#4 in Appendix)

Constituent	Symbol	True period, h	Estimated period, h	Error, %
Lunar	K_1	23.934469	23.933751	0.0030
Lunar	O_1	25.819343	25.819778	0.0017
Lunar	OO_1	22.306075	22.304004	0.0093
Smaller lunar elliptic	M_1	24.833248	24.835465	0.0011
Smaller lunar elliptic	J_1	23.098476	23.097355	0.0049
Larger lunar evectional	ρ	26.723052	26.720821	0.0083
Larger lunar elliptic	Q_1	26.868356	26.868719	0.0014
Larger elliptic	$2Q_1$	28.006223	28.000224	0.0215
Solar diurnal	P_1	24.065891	24.061600	0.0178
Solar	S_1	24	24.002688	0.0112
n/a	σ_1	27.848388	27.849723	0.0048

allow one to obtain very accurate estimates of periods of tidal constituents and, consequently, of any other periodic or quasi-periodic component. This statement is correct as long as there is a sufficient amount of reliable data. Good estimates can be obtained with shorter time series, for example, one year of hourly data, and the resolution will still be high but not as high as when the time series contains 10^5 hourly observations.

4.6 Example of AR Analysis in Time and Frequency Domains

Consider the entire process of time series analysis using as an example the annual values of Tripole Index (TPI) for the Interdecadal Pacific Oscillation (Henley et al. 2015). The time series shown in Fig. 4.9 extends from 1854 through 2018 ($N = 165$); it is closely related to other El Niño-Southern Oscillation indices but differs from them in some respects. The data source is taken from the Web site #5 in Appendix to this chapter. The time series does not contain any statistically significant trend, and its behavior allows one to assume, without any further analysis, that it can be treated as a sample of a stationary random process. The test for Gaussianity showed that the probability density function of this time series can be regarded as normal.

The time series has been analyzed in the time domain by fitting to it AR(p) models of orders from $p = 0$ through $p = 16$ (one-tenth of the time series length). Three of the five order selection criteria used in this book have chosen the order $p = 3$:

$$x_t \approx 0.46x_{t-1} - 0.29x_{t-2} + 0.15x_{t-3} + a_t \tag{4.11}$$

The RMS error of all estimated AR coefficients equals to approximately 0.08 so that the coefficients are statistically significant at the confidence level 0.9 used in this book.

The estimates of the mean value and standard deviation are $\bar{x} \approx -0.15$ and $\hat{\sigma}_x \approx 0.61$. The respective confidence intervals for the mean value and variance estimates obtained for the TPI time series expressed with model (4.11) are [−0.25,

Fig. 4.9 Tripole Index, 1854–2018

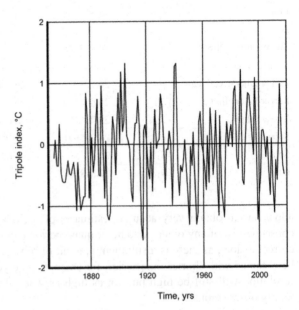

−0.04] and [0.55, 0.68]. These confidence intervals are determined in accordance with Eqs. (4.1)–(4.4) using estimates of the numbers of independent observations $\tilde{N} = 93$ and $\hat{N} = 130$ obtained for the AR(3) model (4.11). These values are calculated through the correlation function estimate under the assumption that the correlation function $r(k)$ at lags $k = 1, 2, 3$ coincides with the sample estimates while its further values behave in the maximum entropy mode. This correlation function obtained according to Eq. (4.5) diminishes very fast so that the numbers of independent observations $\tilde{N}$ and $\hat{N}$ do not differ drastically from the total number of observations N.

The innovation sequence variance $\sigma_a^2 \approx 0.31$ and the predictability (persistence) criterion $r_e(1) = \sqrt{1 - \sigma_a^2/\sigma_x^2}$ equals 0.17 meaning that the unpredictable innovation sequence a_t plays a dominant role in the time series of Tripole Index. This time series is quite close to a white noise sequence, and the variance of its prediction errors will be high.

The characteristic equation of the AR(3) model of TPI given with Eq. (4.11) is

$$1 - 0.46B + 0.29B^2 - 0.15B^3 = 0$$

and it has a pair of complex-conjugated roots $[(-0.05 + 1.81i), (-0.05 - 1.81i)]$ where $i = \sqrt{-1}$. The roots correspond to the natural frequency $f_e \approx 0.25$ cpy. An estimate of the spectrum such as shown in Fig. 4.10a and/or b must be included into analysis of any time series. In this case, the spectrum diminishes with frequency almost monotonically and contains a shelf at frequencies close to 0.25 cpy.

The above described steps are normally required for analysis of any stationary time series.

In concluding this chapter, it should be mentioned that the test for Gaussianity based upon the absolute values of standardized asymmetry and kurtosis is approximate. Thus, the best choice for the TPI time series according to the Kolmogorov–Smirnov criterion is a four-parameter Dagum probability distribution, while the

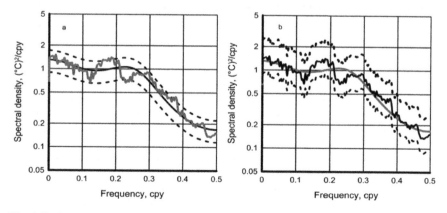

Fig. 4.10 Autoregressive (**a**) and multitaper (**b**) spectral estimates of Tripole Index

normal function is ranked as number three. In this case, the deviation is probably not very important because the best PDF is not too far away from a Gaussian, but it is possible that some distributions that pass this Gaussianity test have properties that will complicate the solution of some problems. This comment may be essential when the values of the process under the study are limited from beneath and/or from the top and often reach their limit(s) within the time interval covered by the time series. If a linear approximation turns out to be unsatisfactory, the first step should probably be to see what PDF is optimal for the time series. An example of an unsatisfactory linear extrapolation is given in Chap. 13 with the sunspot time series. In such cases, one should probably try a nonlinear approach and compare its results with what is obtained with linear methods.

Appendix

1. https://crudata.uea.ac.uk/cru/data/temperature/
2. https://climexp.knmi.nl/data/isunspots.dat
3. http://uhslc.soest.hawaii.edu/data/
4. https://tidesandcurrents.noaa.gov/harcon.html?id=9414317
5. https://www.esrl.noaa.gov/psd/data/timeseries/IPOTPI/.

References

Bendat J, Piersol A (2010) Random data. Analysis and measurements procedures, 4th edn. Wiley, Hoboken

Henley B, Gergis J, Karoly D et al (2015) A tripole index for the interdecadal pacific oscillation. Clim Dyn 45:3077–3090

Thomson R, Emery W (2014) Data analysis methods in physical oceanography. Elsevier, Amsterdam

Yaglom A (1987) Correlation theory of stationary and related random functions. Basic results. Springer, New York

Chapter 5
Stochastic Models and Spectra of Climatic and Related Time Series

Abstract Research activities at all stages of analysis constitute preliminary steps for the most important task—time series forecasting. One of such stages includes efforts to understand statistical properties of the processes that are being studied including probability density functions, spectral densities, and the degree of statistical predictability. Climate is often regarded as a Markov process with a small parameter, which means a slowly and monotonically decreasing spectral density without any oscillations and/or quasi-periodic phenomena. Many climatic time series and indices including AO and AAO, NAO, PDO, AMO, and PNA behave in agreement with that Markov model or even with white noise. The climate indices related to ENSO behave in a different manner: their spectra are nonmonotonic and contain a smooth maximum at about 0.2 cpy. Yet, none of them contains regular oscillations and their predictability stays low. The annual surface temperature for 1920–2018 averaged over large parts of the globe generally does not follow the Markov model, and its predictability is relatively high. Some other oscillatory processes are studied as well, including a version of AAO and MJO—a bivariate random process whose scalar components are shown to possess some statistical predictability.

The final and most important stage of analysis of time series generated by stationary random processes is forecasting. It consists of two parts: determining the achievable quality of forecasting, that is, measuring statistical predictability, and performing the forecast, which means constructing a forecast formula and calculating the future trajectory of the time series with respective confidence intervals.

In order to understand the results of prediction one needs to know what features of the time series have led to its specific forecasts and forecast error variances. This information is implicitly contained in the most essential statistical moment of any stationary time series: its spectral density. If the prediction error variance at the unit lead time coincides with the time series variance, the spectral density will be independent of frequency—a white noise. This random process is unpredictable. If the spectrum is concentrated at low frequencies, the predictability improvement occurs due to the ability to forecast long-term variations of the time series at relatively short lead times. If the spectrum contains a peak whose area composes a significant

© Springer Nature Switzerland AG 2021
V. Privalsky, *Time Series Analysis in Climatology and Related Sciences*,
Progress in Geophysics, https://doi.org/10.1007/978-3-030-58055-1_5

part of the total area under the spectral density curve, the better predictability occurs due to the presence of a quasi-periodic or cyclic, component.

Thus, time series analysis must provide a time domain model of the time series and an estimate of the spectral density corresponding to it. This chapter contains results of analysis of time series obtained from observations, mostly at climatic time scales. Specifically, the tasks here are to describe typical time domain models of different types of climatic time series and to characterize the behavior of their spectral densities.

Essentially, Sects. 5.1 and 5.2 can be regarded as an attempt to sum up information about the typical behavior of climate as a stationary random process starting from its first stochastic model suggested by Hasselmann (1976) in the form of a Markov process. The early estimates of climate spectra used in particular, to verify the model, include publications by Privalsky (1976, 1977) and by Frankingnoul and Hasselmann (1977). The other goal is to see if the concept of low statistical predictability of respective climate models agrees with observation data. The task of time series forecasting will be discussed and illustrated with examples in Chap. 6.

5.1 Properties of Climate Indices

At climatic time scales, the basic statistical properties of a large number of geophysical time series—about 3000—has been summarized in the fundamental work by Dobrovolski (2000) dealing with stochastic models of scalar climatic data. The time series in that book include surface temperature, atmospheric pressure, precipitation, sea level, and some other geophysical variables observed at individual stations; the data set includes 195 time series of sea surface temperature averaged within $5° \times 5°$ squares. Most of those time series are best approximated with either a white noise or a Markov process (Dobrovolski 2000, p. 135). The white noise model AR(0) can be justly regarded as a specific case of the AR(1) model. The prevalence of the AR(1) model for climatic time series obtained without large-scale spatial averaging has been noted recently in Privalsky and Yushkov (2018), but the results given in Dobrovolski (2000) are based upon a much larger observation base.

In this section, we will complement the available information by studying first a number of geophysical time series that are often used as climate indicators or indices; their names usually contain the term "oscillation" or "index." The list is given in Table 5.1, and the data sources are shown in the Appendix to this chapter. In all cases, the value of Δt is one year. Along with the optimal AR orders p for the time series, the table contains the values of statistical predictability criterion (3.7): $r_e(1) = \sqrt{1 - \sigma_a^2/\sigma_x^2}$, where σ_x^2 and σ_a^2 are the time series variance and the variance of its innovation sequence.

The sources of data listed in the table are given in Appendix to this chapter: the numbers in the first column of the table coincide with the numbers in the Appendix.

Two characteristic features are common for the time series in Table 5.1: all of them can be regarded as Gaussian, and, with one exception, all of them have low

Table 5.1 Basic properties of climate indices, $\Delta t = 1$ year

##	Name	Years	N, yrs	p	$r_e(1)$	Trend after 1980
1	Antarctic Oscillation (AAO)	1957–2018	62	0	0	Yes
2	Atlantic Multidecadal Oscillation (AMO)	1854–2018	165	2	0.62	No[a]
3	Arctic Oscillation (AO)	1899–2017	119	0	0	Yes
4	Dipole Mode Index (DMI)	1870–2017	148	0	0	No
5	ENSO atmospheric component (SOI)	1866–2018	153	2	0.20	No
6	ENSO oceanic component (NINO3.4)	1870–2018	149	2	0.35	No
7	ENSO, precipitation index (GPCC)	1891–2018	128	2	0.35	No
8	ENSO, Modoki El Niño Index (MEI)	1950–2016	67	2	0.36	No
9	North Atlantic Oscillation (NAO)	1865–2001	137	0	0	No
10	North Pacific Index (NPI)	1959–2017	59	0	0	No
11	Pacific Decadal Oscillation (PDO)	1854–2018	165	1	0.46	No
12	Pacific-North American Pattern (PNA)	1851–2014	164	1	0.41	Yes
13	Interdecadal Pacific Oscillation Tripole Index (TPI)	1870–2018	149	3	0.44	No

[a]Since about 1997

statistical predictability. This means that they present sample records of random processes similar to a white noise; that is, their behavior in the time domain is very irregular, and, consequently, none of them contains oscillations as the term is understood in physics. The exception is the relatively high predictability of the Atlantic Multidecadal Oscillation. Thus, judging by the low optimal AR orders and the low predictability, one may say that though the optimal model for most of these time series is not AR(1) their behavior does not contradict the assumption of the Markov character of climate variability and that the value of the autoregressive coefficient is significantly smaller than one.

Most of the 13 climate indices in Table 5.1 do not show any response to the global warming in the form of a trend; the reason for this feature is not clear. In seven cases, the names include the term "oscillation," that is, regularly repeating deviations from some equilibrity level. Visually, none of those time series contains oscillations. The presence of oscillations can also be seen from the spectral density and/or from the roots of the characteristic equation corresponding to a given AR model: at least, some of the roots must be complex-valued and the damping coefficient should not be too large. For example, the Antarctic Oscillation (or the Antarctic Oscillation Index) is defined as the difference of mean zonal atmospheric pressure at sea level between 40° S and 65° S and can be regarded as similar to NAO (Gong and Wang 1999). The optimal model for both NAO and AAO time series is a white noise, that is, a sequence of identically distributed and mutually independent random variables; it cannot contain oscillations and hardly carries any useful probabilistic information. The AR orders 0 and 1 exclude the presence of oscillations in respective time series because their spectral densities are frequency independent when $p = 0$ or decrease monotonically with growing frequency when $p = 1$. Oscillations may exist in time series whose AR orders exceed 1. In this case, there are six such indices: AMO, SOI, NINO3.4, GPCC, MEI, and TPI. (This TPI time series differs from the time series analyzed in Chap. 4.)

The appearance of the term "oscillation" in the case of the Atlantic Multidecadal Oscillation described with an AR(2) model is probably explained by the presence of a couple of long-term (at least 60 years) quasi-oscillatory variations in the AMO graph (Fig. 5.1a). This is not sufficient to believe that AMO behaves in an oscillatory manner; actually, AMO does not contain any regular oscillations, its spectrum does not show any peaks (Fig. 5.1b), and the roots of its characteristic equation with the AR coefficients $\varphi_1 \approx 0.51$ and $\varphi_2 \approx 0.17$ are real.

The situation with possible presence of oscillations in climate indices becomes different for the five time series directly related to ENSO: SOI, NINO3.4, GPCC, MEI, and TPI. Having in mind the belief that ENSO plays an important role of in the Earth's climate prevalent in climatology, the ENSO case deserves a more detailed analysis.

In all five ENSO-related cases, the characteristic equations have complexly valued roots, which means that the term "oscillation" is proper for SOI and NINO3.4 as well as for GPCC, MEI, and TPI. However, the values of the damping coefficients d lie between 0.43 year^{-1} (SOI) and 0.55 year^{-1} (MEI). Due to this high damping rate, the

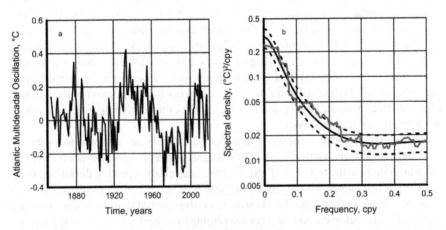

Fig. 5.1 AMO, 1854-2018 (**a**); MEM (gray) and MTM (black) spectral density estimates (**b**)

oscillations at respective natural frequencies dissipate very quickly, cannot be clearly seen in the time series and create rather smooth maxima in the spectral estimates.

The time series of SOI, NINO3.4, GPCC, and MEI are best described with AR(2) models, that is,

$$x_t = \varphi_1 x_{t-1} + \varphi_2 x_{t-1} + a_t \qquad (5.1)$$

and in all four cases the AR coefficients are statistically significant with $\varphi_1 > 0$ and $\varphi_2 < 0$. For example, the coefficients are 0.30 and -0.28 for NINO3.4 and 0.18 and -0.18 for SOI. The roots of respective characteristic equations are complex-valued and correspond to natural frequencies f_e of 0.25 cpy (MEI), 0.20 cpy (GPCC and NINO3.4), and 0.22 cpy (SOI). The AR(3) model for the TPI time series also has complex-valued roots with the natural frequency of about 0.25 cpy.

With these large damping coefficients, the oscillations disappear as soon as they emerge but their existence affects the spectra of all five time series which constitute the ENSO system (SOI, NINO3.4) or are closely related to it (GPCC, MEI, TPI).

The spectral density estimates of the four ENSO-related time series do not contain any definite sharp peaks anywhere between 0.18 cpy and 0.24 cpy, but the shapes of all those spectra estimated through respective AR(2) models are quite similar to each other (Fig. 5.2). The spectrum of TPI contains higher energy at lower frequencies and a shelf at about 0.20 cpy. It behaves similar to what is shown in Fig. 4.10 for a different version of TPI.

The bell-shaped spectra of ENSO-related time series seem to be rather unique in climatology. Theoretically, a nonmonotonic change of the spectral density means that respective time series may have some significant predictability. However, as seen from Table 5.1, in all five oscillatory cases of ENSO-related climate indices the predictability criterion $r_e(1)$ is low staying between 0.20 (SOI) and 0.44 (TPI).

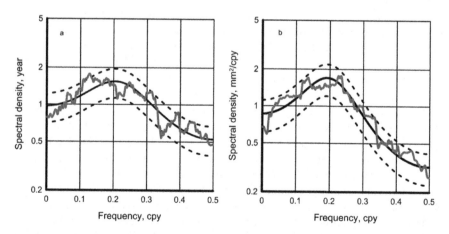

Fig. 5.2 Estimated spectral densities of oscillatory ENSO-related time series: SOI (**a**) and GPCC (**b**). The MTM estimates are shown in gray

Consequently, one may say that at climatic time scales all ENSO-related indices behave in a manner which is close to a white noise. They do not contain any regular oscillations, and their probabilistic predictions at $\Delta t = 1$ year would hardly produce satisfactory results.

5.2   Properties of Time Series of Spatially Averaged Surface Temperature

With the exception of ENSO-related phenomena, the results of analysis of geophysical time series listed in Table 5.1 do not contradict the hypothesis of the Markovian behavior of climate (Hasselmann 1976). Out of the 13 time series in Table 5.1, seven have orders not higher than 1, which can be regarded as a confirmation of the hypothesis. The other six samples have low predictability, which does not differ much from predictability of the remaining seven time series. The predictability of AMO is better than in all other cases, and it may be high enough for practical applications. The AMO time series differs from other time series in the table in the sense that it is obtained by averaging SST over a large area of the North Atlantic; therefore, one can assume that the comparatively high rate of spectral density decrease and the higher predictability criterion $r_e(1) \approx 0.62$ for AMO could be the result of that averaging.

The global climate is better characterized with data obtained by averaging over large parts of the globe. The AMO time series is just a specific example of such averaging, but we have nine time series that show the surface temperature over the entire globe, its hemispheres, and oceanic and terrestrial parts. Those time series have been analyzed in Privalsky and Yushkov (2018) and found to have a more complicated structure and a higher predictability than the other time series studied in that work.

The data used in the above publication include the complete time series given by the University of East Anglia; most of the time series begin in 1850. The authors of the data files show that the degree of coverage during the XIX Century was poor. Following the example given in Dobrovolski (2000), we will study the same time series starting from 1920, when the coverage with observations generally increases to 50% and higher for the global, hemispheric and oceanic data.

The results given in Table 5.2 confirm one of the previous conclusions: the annual surface temperature averaged over large parts of the globe is best described with relatively complicated models having AR orders $p = 3$ or $p = 4$ and a relatively high statistical predictability. The results for the southern hemisphere as a whole and for its land follow a Markov model and have lower statistical predictability; they agree with our results obtained from the data given by the Goddard Institute of Space Studies (GISS). According to the GISS data for the southern hemisphere (#14 in Appendix), the autoregressive order $p = 1$ and the criterion $r_e(1) \approx 0.55$.

The data sets show that spatial averaging on the global scale and over the northern hemisphere including its oceans and land produces time series whose properties

Table 5.2 Basic properties of spatially averaged and detrended temperature data, 1920–2018

Name	p	$r_e(1)$	Name	p	$r_e(1)$
Global	4	0.71	Ocean, southern hemisphere	3	0.57
Northern hemisphere	4	0.77	Land	4	0.74
Southern hemisphere	1	0.44	Land, northern hemisphere	4	0.76
Ocean	3	0.71	Land, southern hemisphere	1	0.40
Ocean, northern hemisphere	3	0.82			

differ quite significantly from what is shown in Table 5.1 for individual climate indices. The optimal AR orders increase up to four, and the predictability criterion grows up to 0.82 for the north hemispheric ocean. The reason for the behavior of temperature over the southern hemisphere for the time series which begin in 1920 is not clear, but it may be related to the change is statistical properties of the trivariate system consisting of the time series of global, land, and terrestrial time series. For example, the predictability criterion $r_e(1)$ for the entire time series is 0.74 (Privalsky and Yushkov 2018) and 0.44 for the time series that begins in 1920. A more detailed description of the change is given in Chap. 14.

The predictability criterion used by Hasselmann (1976) in his Eq. (6.3) can be given as

$$s^2 = \delta^2/(\varepsilon^2 + \delta^2), \tag{5.2}$$

where δ^2 and ε^2 are the variances of the predictable and unpredictable parts of the time series (signal and noise, according to K. Hasselmann). In our notations, ε^2 coincides with σ_a^2 and $\delta^2 = \sigma_x^2 - \sigma_a^2$. Therefore, $s^2 = r_e^2(1)$. In Hasselmann's opinion, the statistical predictability criterion for stationary climate systems generally does not exceed 0.5 and actually is always much less than unity. The results in Table 5.1 agree with that statement but spatial averaging seems to lead to more complicated models and to better statistical predictability (Table 5.2). All these time series successfully pass the test for Gaussianity.

Thus, the predictability criterion $r_e(1)$, which also coincides with the correlation coefficient between the current value of the time series and its extrapolation at the unit lead time, increases to 0.82 for the annual ocean surface temperature averaged over the northern hemisphere. A stationary time series with such statistical predictability may be forecasted with relatively small error variance for several lead times (years). Obviously, the predictability will be even higher if the trend in the temperature is caused by natural factors (see Chap. 6).

Thus, the simple Markov model, which is good for many climate processes, is not necessarily optimal for the time series of annual surface temperature obtained through spatial averaging of data over the globe and over its major parts, with the exception of the southern hemisphere and its land where the data seem to be less reliable. The averaging leads to more complicated models and to some improvements in statistical predictability, at least in the cases of the global and northern hemispheric temperature.

5.3 Quasi-Biennial Oscillation

The "rule of no significant sharp peaks" in climate spectra has at least one exception which is supported with decades of direct observations. At least one atmospheric process—the Quasi-Biennial Oscillation, or QBO—does not follow this rule. The QBO phenomenon exists in the equatorial stratosphere at altitudes from about 16 km to 50 km, and it is characterized with quasi-periodic variations of the westerly and easterly wind speed. The period of oscillations is about 28 months, which corresponds to the frequency of about 0.43 cpy. It has been discovered in the 1950's and investigated in a number of publications, in particular, in Holton and Lindzen (1972) who proposed a physical model for QBO. In the review of QBO research by Baldwin et al (2001), QBO is called "a fascinating example of a coherent, oscillating mean flow that is driven with propagating waves with periods unrelated to the resulting oscillation." Some effects of QBO upon climate are discussed by Anstey and Shepherd (2014).

The statistical properties of QBO such as its spectra and statistical predictability do not seem to have been analyzed within the framework of theory of random processes; this section (along with Chaps. 6 and 10) is supposed to fill this gap in the part related to QBO as a scalar and bivariate (Chap. 10) phenomenon. It will be analyzed here using the set of monthly observational data provided by the Institute of Meteorology of the Free University of Berlin for the time interval from 1953 through December 2018 (see #15 in Appendix and Naujokat 1986). The set includes monthly wind speed data in the equatorial stratosphere at seven atmospheric pressure levels, from 10 to 70 hPa; these levels correspond to altitudes from 31 km to 18 km.

If the goal of the study were to analyze QBO as a scalar random process, the data could have been taken at the sampling interval $\Delta t = 6$ months or even 1 year. As QBO's statistical predictability at a monthly sampling rate will also be studied in Chap. 6, the sampling interval $\Delta t = 1$ month is taken in this section as well. Examples of QBO variations are shown in Fig. 5.3.

The basic statistical characteristics of QBO are shown in Table 5.3. The average wind speed is easterly (negative), and it decreases below the 20 hPa level turning eastward at the lowest level. The variance increases from the 10 hPa level by about 10% to 15 hPa and 20 hPa and then gradually decreases downward by an order of magnitude. These facts are well known (e.g., Baldwin et al. 2001). The optimal AR models have orders from $p = 11$ to $p = 29$; such orders are too high for individual time domain analysis.

The typical shape of the spectrum shows an almost periodic random function of time at $f \approx 0.43$ cpy (Fig. 5.4a). The maximum is very narrow and completely dominates the spectrum so that a more detailed picture can only be seen when the scale is logarithmic along both axes (Fig. 5.4b). This seems to be an absolutely

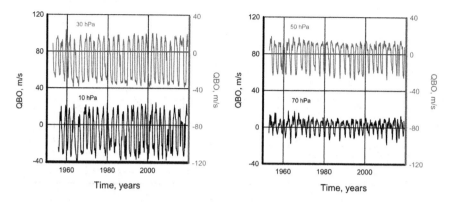

Fig. 5.3 Examples of monthly wind velocity data at different atmospheric pressure levels

Table 5.3 Basic statistical characteristics of the QBO data

Level, hPa	Altitude, km	Average wind speed, m/s	Variance, (m/s)^2	AR order, p	f_e, cpy
10	31.1	−7.8	357	11	0.432
15	28.4	−8.4	404	29	0.432
20	26.5	−8.3	391	11	0.432
30	23.8	−6.4	326	13	0.432
40	22.0	−2.9	252	16	0.432
50	20.6	−0.1	167	23	0.432
70	18.4	1.9	42	23	0.432

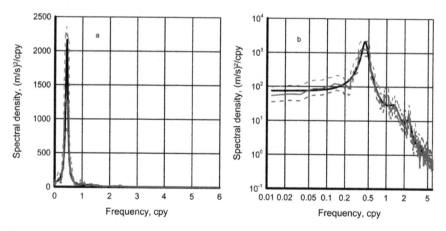

Fig. 5.4 Typical shape of QBO spectral density: MTM (gray) and MEM (black) estimates (10 hPa level)

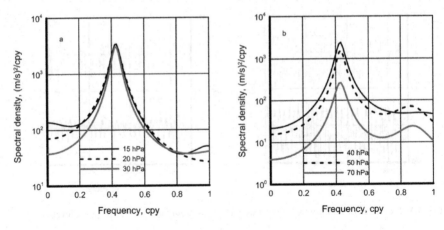

Fig. 5.5 QBO spectra at levels from 15 hPa to 70 hPa

unique phenomenon at climatic time scales. At higher frequencies, the spectral density diminishes rather quickly with all other peaks being statistically insignificant. Having this in mind, the spectra will be shown in what follows at frequencies not exceeding 1 cpy.

The AR spectral estimates of QBO at all other levels are given in Fig. 5.5. In all seven cases (including the 10 Pa level in Fig. 5.4), the frequency of the major spectral peak is found at $f = 0.432$ cpy. A tenfold increase in the frequency resolution from 0.012 cpy to 0.0012 cpy showed that the period of oscillation determined through the frequency of the maximum spectral density in all seven estimates varies by less than 2% and stays between 27.7 and 28.1 months. The average over the seven spectral estimates frequency of the spectral peak corresponds to the average period equal to 27.99 month. This very clear-cut stability can be regarded as another distinctive property of the QBO phenomenon.

According to the model suggested by Holton and Lindzen (1972), the Quasi-Biennial Oscillation is forced by vertically propagating planetary waves with periods of 5–15 days. This explanation means that at altitudes where QBO occurs the equatorial stratosphere works as a filter that transforms the upward propagation of this high-frequency noise into a downward moving oscillation whose frequency is up to two orders of magnitude lower that the frequency of the forcing.

As seen from these descriptions and from the figures, the QBO system has a complicated structure and it should be studied as a multivariate random process. This will be done in Chap. 10.

5.4   Other Oscillations

The Madden–Julian Oscillation (MJO) is another unusual phenomenon both because it is not firmly fixed geographically and because it presents an oscillatory system not related to tides or to a seasonal trend. A review of MJO can be found in Zhang (2005).

Strictly speaking, the MJO phenomenon is a vector process and its spectra should be estimated in agreement with the approach discussed in Thomson and Emery (2014, Chap. 5). However, having in mind the methodological goals of the book, the MJO components will be treated here as either two scalar time series (this chapter and Chap. 6) or as a bivariate process (Chap. 8).

The MJO data used here consist of daily MJO indices RMM1 and RMM2 from January 1, 1979 through April 30, 2017 ($N = 14000$, $\Delta t = 1$ day). Thus, MJO is a bivariate random process. The source of the data is the Australian Bureau of Meteorology, site #16 in Appendix. The graph of the time series is shown in Fig. 5.6a. The hypothesis of stationarity can be accepted through visual assessment, but it is also confirmed by using the method described in Chap. 4. The spectral densities of the time series components are very similar and contain a single wide peak at the frequency close to 0.02 cpd. The spectral estimates are shown in Fig. 5.6b for the part of the frequency axis up to 0.05 cpd; at higher frequencies, the spectrum is monotonically decreasing. The confidence limits are not shown because they almost coincide with the spectra due to the high reliability of estimates obtained with these long time series. The contribution of higher frequencies is negligibly small. Thus, the Madden–Julian Oscillation presents a good example of an oscillatory system. The statistical predictability criterion $r_e(1)$ given with Eq. (3.7) amounts to about 0.98, meaning that both components possess high statistical predictability at the unit lead time, that is, at 1 day.

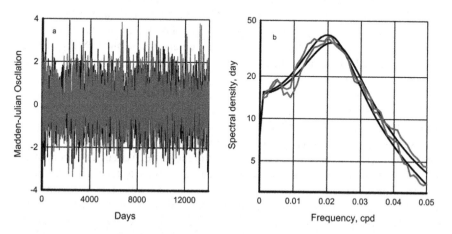

Fig. 5.6 Components RMM1 (black) and RMM2 (gray) of Madden–Julian Oscillation (**a**) and their MEM (smooth) and MTM (rugged) spectral estimates (**b**)

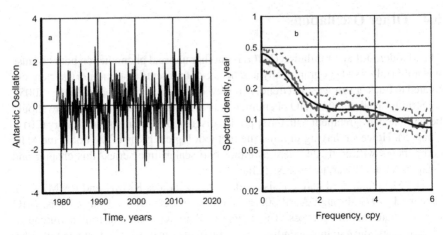

Fig. 5.7 Antarctic Oscillation (**a**) and its MTM (gray) and MEM (black) spectral estimates (**b**)

Another supposedly oscillatory system is the Antarctic Oscillation—an index based upon zonally averaged sea level pressure between 40° S and 65° S (e.g., Gong and Wang, 1999). A set of monthly data on the Antarctic Oscillation Index (AAO), 1979-2018, $N = 480$, $\Delta t = 1$ month, is available at site #17 in Appendix (Fig. 5.7a). This is not the same time series as the one given in Table 5.1. The optimal AR model for AAO is AR(3) with no peaks in its spectral density estimate (Fig. 5.7b). The multitaper spectral analysis according to Thomson's method agrees with the AR (or MEM) result.

Some other estimates of AAO spectra and even the coherence function (see Chap. 7) relating AAO to other phenomena (e.g., Pohl et al. 2010; Artamonov et al. 2016) show low-frequency peaks in the spectra and coherence, but the estimates are given there without confidence intervals, which makes them untrustworthy.

The list of time series representing other systems, which are arguably called oscillatory, can be extended to include, among others, the Summer North Atlantic Oscillation and the Indian Ocean Dipole with its two components; in all those cases, the properly built estimates of spectral density show the lack of oscillations.

The results of analysis given in this chapter and previously in Privalsky and Yushkov (2018) indicate that, with the exceptions of QBO and, to a lesser degree, ENSO, the geophysical processes at climatic scales have monotonic or almost monotonic spectra. The models become more complicated and their predictability increases when the data are averaged over the large parts of the globe. Actually, this conclusion about the role of spatial averaging had been made quite a while ago (Privalsky 1983).

The typical spectrum of ENSO-related processes contains a wide maximum at about 0.2 cpy, but all those phenomena remain quite close to a white noise. The seasonal and daily trends in geophysical phenomena (possible cases of a periodically correlated random processes) are not studied here, but the Madden–Julian Oscillation with its meteorological time scales is a cyclic phenomenon which may have some significant predictability. At the same time, one should remember that the analysis of MJO is given here and in Chaps. 6 and 8 to illustrate the process of time series analysis.

Appendix

1. https://www.cpc.ncep.noaa.gov/products/precip/CWlink/daily_ao_index/aao/monthly.aao.index.b79.current.ascii
2. https://climexp.knmi.nl/data/iamo_ersst_ts.dat
3. https://climexp.knmi.nl/data/iao_slp_ext.dat
4. https://www.esrl.noaa.gov/psd/gcos_wgsp/Timeseries/DMI
5. https://crudata.uea.ac.uk/cru/data/soi/soi.dat
6. https://climexp.knmi.nl/data/ihadisst1_nino3.4a.dat
7. https://climexp.knmi.nl/data/itelecon_nino34_gpcc.dat
8. https://climexp.knmi.nl/data/imei.dat
9. https://climexp.knmi.nl/data/inao_ijs_azo.dat
10. https://ds.data.jma.go.jp/tcc/tcc/products/elnino/decadal/npiwin.txt
11. https://www.ncdc.noaa.gov/teleconnections/pdo/
12. https://www.esrl.noaa.gov/psd/data/20thC_Rean/timeseries/monthly/PNA/pna.20crv2c.long.data
13. https://www.esrl.noaa.gov/psd/data/timeseries/IPOTPI/tpi.timeseries.hadisst11.data
14. https://data.giss.nasa.gov/gistemp/tabledata_v4/SH.Ts+dSST.txt
15. https://www.geo.fu-berlin.de/met/ag/strat/produkte/qbo/qbo.dat
16. http://www.bom.gov.au/climate/mjo/graphics/rmm.74toRealtime.txt
17. https://climexp.knmi.nl/data/icpc_aao.dat.

References

Anstey J, Shepherd T (2014) High-latitude influence of the quasi-biennial oscillation. Q J Roy Meteor Soc 140:1–21
Artamonov J, Fedirko A, Skripaleva E (2016) Climatic variability of transport in the upper layer of the antarctic circumpolar current from hydrological and satellite data. Izv AN Fiz Atmos Ok 52:1051–1063
Baldwin M, Gray L, Dunkerton T et al (2001) Quasi-biennial oscillation. Rev Geophys 29:179–329
Dobrovolski S (2000) Stochastic climate theory. Models and applications. Springer, Berlin
Frankingnoul C, Hasselmann K (1977) Stochastic climate models. Part II. Applications to sea surface temperature anomalies and thermocline variability. Tellus 29:359–370
Gong D, Wang S (1999) Definition of Antarctic oscillation index. Geophys Res Lett 26:459–462
Hasselmann K (1976) Stochastic climate models. Part I. Theory Tellus 28:473–485
Holton J, Lindzen R (1972) An updated theory for the Quasi-Biennial Cycle of the tropical stratosphere. J Atmos Sci 29:1076–1080

Naujokat B (1986) An update of the observed Quasi-Biennial Oscillation of the stratospheric winds over the tropics. JAS 43:1873–1877

Pohl B, Faucereau N et al (2010) Relationships between the Antarctic Oscillation, the Madden-Julian Oscillation, and ENSO, and consequences for rainfall analysis. J. Climate 23:238–255

Privalsky V (1976) Estimating the spectral densities of large-scale processes. Izv. Atmos Ocean Phys 12:979–982

Privalsky V (1977) The statistical predictability of the large-scale hydrometeorological processes. Izv. Atmos Ocean Phys 13:261–265

Privalsky V (1983) Statistical predictability and spectra of air temperature over the northern hemisphere. Tellus 35A:51–59

Privalsky V, Yushkov V (2018) Getting it right matters: climate spectra and their estimation. Pure Appl Geoph 175:3085–3096

Thomson R, Emery J (2014) Data analysis methods in physical oceanography, 3rd edn. Elsevier, Amsterdam

Zhang C (2005) Madden-Julian Oscillation. Rev Geophys 43:1–36

Chapter 6
Statistical Forecasting of Geophysical Processes

Abstract The Kolmogorov–Wiener theory of extrapolation of stationary random processes (KWT) has been created about 80 years ago. The theory proves that no linear method of extrapolation (prediction, forecasting) can be better than what is possible to achieve through KWT. If the time series is Gaussian (which happens in climatology and geophysics in general), the KWT is the best possible solution for time series forecasting, be it linear or nonlinear. Any method of statistical forecasting must be tested against KWT which is never done in geophysics and solar research because KWT is not known in Earth and solar sciences. This chapter contains a brief description of KWT, a proof of its efficiency, and examples of its application with climatic and meteorological time series having different predictability properties. The examples include the annual global surface temperature, QBO, oceanic and atmospheric components of ENSO, and MJO. The high efficiency of KWT is demonstrated with QBO and MJO, while the extrapolation of global annual temperature may be acceptable for up to 5–7 years if its trend is regarded as a nature-caused factor. The example with the oceanic component of ENSO is successful for at least eight months. Variations of SOI are unpredictable.

6.1 General Remarks

Forecasting geophysical processes is probably the most desired goal in Earth and related solar sciences. Reliable predictions are needed at time scales from hours and days (meteorology, hydrology, etc.) to decades and centuries (climatology and related sciences). With one exception, all geophysical processes in the atmosphere—ocean—land—cryosphere system are random, which means that none of them can be predicted at any lead time without an error. The exception is tides—a deterministic process which exists in the oceans, atmosphere, and in the solid body of the planet. The knowledge of tides is especially important for the oceans, and tides in the open ocean can be predicted almost precisely. Along the shorelines where tides play an important role, sea level variations can generally be predicted with sufficient accuracy as well, but there may be some cases when random disturbances should also be taken into account (Munk and Cartwright 1966).

© Springer Nature Switzerland AG 2021
V. Privalsky, *Time Series Analysis in Climatology and Related Sciences*,
Progress in Geophysics, https://doi.org/10.1007/978-3-030-58055-1_6

The behavior of another astronomically caused process—the seasonal trend—is so irregular that one cannot even say for sure whether the next summer (or any other season) will be warmer or cooler than the current one.

The atmospheric, oceanic, terrestrial, and cryospheric processes and their interactions can be described with fluid dynamics equations; however, the equations are complicated, numerous, and cannot be solved analytically. Getting reliable numerical solutions encounters serious physical and computational problems, which cannot be discussed in this book. However, there is at least one important example of successful numerical solution of prediction problems—the weather forecasting. The forecasts given by meteorologists are reliable and rarely contain serious errors at lead times at least up to about a week. These forecasts are obtained by uploading information about the current (initial) state of processes involved in weather generation into a numerical computational scheme having discrete temporal and spatial resolution and then running the scheme forward in time and space to obtain forecasts. As the knowledge of the initial conditions cannot be ideal, the forecasts contain errors. Besides, the computational grid is discrete so that the processes whose scales are smaller than the distance between the grid nodes and shorter than the unit time step cannot be directly taken into account. The errors in the initial and other conditions grow with the forecast lead time, and eventually, the variance of the forecast errors becomes equal to the variance of the process that is being forecasted. The forecast becomes unusable. It means that the process has a predictability limit; the limit should be defined quantitatively through the ratio of the forecast error variance as a function of lead time to the variance of the process. These issues have been discussed in a number of classical works by Lorenz (1963, 1975, 1995).

For weather forecasting, the predictability limit at which the error variance approaches the variance of the process amounts to about a week or slightly longer. The numerical models of climate used, first of all, to assess the influence of anthropogenic factors upon future climate require the same equations and initial conditions as in weather forecasting; however, the temporal and spatial resolutions of numerical climate models are much less detailed and the models cannot predict the natural variability of climate. This may be the reason (or one of the reasons) why the results of climate simulations with numerical general circulation models that show the behavior of climate in the twenty-first century are called climate projections rather than climate predictions (IPCC 2013).

Thus, the numerical models cannot ensure predictions beyond the predictability limits defined in accordance with Lorenz's ideas. Under this situation, it becomes quite reasonable to deal with the problem by trying to predict the state of Earth system's elements at lead times of weeks, months, or even longer by using the probabilistic approach (also see Lorenz 2007). This is the subject discussed in this chapter: probabilistic (traditionally—statistical) extrapolation of geophysical time series based upon information about their behavior in the past. The term "extrapolation" is equivalent to prediction and forecasting; for example, forecasting a geophysical process by its behavior in the past and, possibly, by the past behavior of other predictors is nothing else but an extrapolation of a random process.

Fortunately, the problem of extrapolation of stationary random processes represented with time series had been solved independently of each other by two giants of twentieth-century mathematics: the founder of probability theory as a mathematical discipline Andrey Kolmogorov and the creator of cybernetics Norbert Wiener. A. Kolmogorov published two articles which describe his theory of linear interpolation and extrapolation of discrete random sequences (Kolmogorov 1939, 1941). At about the same time, N. Wiener developed a more general theory of extrapolation, interpolation, and smoothing of discrete and continuous stationary random processes with the same criterion of extrapolation efficiency as in the Kolmogorov's work—the variance of extrapolation error.

The goal of Wiener's research was to solve an engineering problem (Wiener 1949), and it resulted in a practical method of extrapolation as a part of efforts to operate anti-aircraft artillery fire. The method suggested by N. Wiener was rather cumbersome and the first (classified) edition of his book on the subject printed in 1942 in a yellow binding became known as "the Yellow Peril" (see Coales and Kahne 2014). Slightly simplifying the theoretical results obtained by A. Kolmogorov and N. Wiener, one may say that if the integral of the spectral density logarithm is larger than $-\infty$, the stationary random function that is to be extrapolated will be linearly regular and may possess some predictability (another example of the key role played by the spectral density).

In 1952, A. Yaglom published his classical work on theory of stationary random functions, which contains, in particular, a simpler approach to a practical solution of the extrapolation task for an important class of stationary random processes. An English translation has been published a decade later (Yaglom 1962). The spectral characteristic for extrapolation of stationary random sequences and processes introduced by A. Yaglom made the solution of the task much simpler. Later, M. Fortus obtained an explicit expression for the extrapolation error variance in the case when the spectral density presents a rational function of frequency (Fortus 1978). Such discrete processes are also called the autoregressive and moving average (ARMA) random sequences.

The degree of predictability of any stationary random sequence is characterized with its spectral density in the scalar case and the spectral matrix in the multivariate case. The concept of predictability is related to the lead time at which the error variance remains sufficiently smaller than the variance of the process that is being predicted. A quantitative criterion can be the ratio of the prediction error variance to the variance of the process (or its complement to unity as in previous chapters) while the shape of the spectral density provides a good qualitative criterion. If the spectrum is concentrated within a narrow frequency band, the predictability will be relatively high. In geophysics, it can be a quasi-periodical process such as the Quasi-Biennial Oscillation or a process whose spectrum is concentrated at low frequencies, such as the annual global surface temperature. If the spectrum is distributed more or less evenly over the frequency axis, the process has low statistical predictability. These are some of the reasons why the knowledge of the spectrum is necessary for understanding a random process and for working with its sample records, that is, with time series.

Further developments in extrapolation theory, including the case of multivariate processes and the generalized case when the process contains singular components such as strictly periodic functions are discussed in Yaglom (1962); important earlier theoretical publications on extrapolation include Zasukhin (1941) and Zade and Ragazzini (1950).

The practical solution of the linear extrapolation task became much easier with the publication in 1970 of the classical book by George E. P. Box and Gwilym M. Jenkins. Between 1970 and 2015, the book has been updated and published four more times (from Box and Jenkins 1970 to Box et al. 2015).

The requirement of minimizing the variance of linear extrapolation error that lays in the basis of the Kolmogorov–Wiener theory (KWT) becomes all-embracing in the case of Gaussian processes. Namely, if a stationary random sequence is Gaussian, "the best linear extrapolation formula is automatically the best *possible* extrapolation formula" (Yaglom 1962, p. 99; italics by A. Yaglom).

Many geophysical processes can be treated as Gaussian, in particular, at climatic scales, which means that in geophysics no method of extrapolation (forecasting, prediction) of Gaussian stationary random process, scalar or multivariate, can be better than the extrapolation within the framework of the linear Kolmogorov–Wiener theory. It also means that irrespective of the shape of the probability density function, there may be no linear extrapolation methods with the error variance smaller than what follows from KWT. According to Yaglom (1962, p. 100), "… in many cases, the mean square error of the best linear extrapolation only slightly exceeds the mean square error of the best nonlinear extrapolation." Thus, the results obtained by applying the Kolmogorov–Wiener theory of extrapolation to scalar stationary time series are either the best possible or close to the best possible results. The statement is also true for multivariate stationary processes.

This property of KWT means that whenever a time series generated by a Gaussian stationary random process, scalar or multivariate, is being forecasted, there is one and only one proper solution: forecasting it within the framework of the Kolmogorov–Wiener theory of extrapolation. If the process is not Gaussian, a smaller error variance can be obtained only through nonlinear forecasting. There is no doubt that processes of this type exist; in such cases, it is necessary to compare the method of extrapolation with a KWT method to show the advantages of the selected nonlinear approach.

6.2 Method of Extrapolation

In both scalar and multivariate cases, the extrapolation means a forecast of the time series on the basis of its behavior in the past. The method of extrapolation used in this book to predict the behavior of stationary geophysical time series is based upon the autoregressive modeling (Box et al. 2015). It is discussed in this chapter for the case of scalar time series x_t known over a finite time interval from $t = \Delta t$ through $t = N \Delta t$. The sampling interval Δt is the unit time step, which can be a minute, hour, month, year, or whatever the data prescribes. Here, $\Delta t = 1$. The only assumption made about the time series x_t is that it presents a sample record of a stationary random process.

The first stage of extrapolation procedure is to approximate the scalar time series with an AR model of a properly selected order p. The result of approximation is

$$x_t = \varphi_1 x_{t-1} + \cdots + \varphi_p x_{t-p} + a_t, \tag{6.1}$$

where $\varphi_j, j = 1, \ldots, p$ are the AR coefficients and a_t is a zero mean innovation sequence (white noise) with the variance σ_a^2.

Equation (6.1) describes the time series as a function of its behavior in the past, that is, exactly what is required for time series extrapolation. The unknown true value of the time series at lead time τ is

$$x_{t+\tau} = \varphi_1 x_{t+\tau-1} + \cdots + \varphi_p x_{t+\tau-p} + a_{t+\tau} \tag{6.2}$$

so that at the lead time $\tau = 1$

$$x_{t+1} = \varphi_1 x_t + \cdots + \varphi_p x_{t-p+1} + a_{t+1} \tag{6.3}$$

At time t, all terms in the right-hand side of this equation, with the exception of a_{t+1}, are known because they belong to the observed initial time series. Therefore, the extrapolated (predicted, forecasted) value of the time series at the unit lead time is

$$\hat{x}_t(1) = \varphi_1 x_t + \cdots + \varphi_p x_{t-p+1}. \tag{6.4}$$

As the extrapolation error at the unit lead time is a_{t+1}, its variance is σ_a^2. For $\tau = 2$, one has

$$\hat{x}_t(2) = \varphi_1 \hat{x}_t(1) + \cdots + \varphi_p x_{t-p+2} \tag{6.5}$$

so that the extrapolation error will be the sum of σ_a^2 with the error at $\tau = 1$ (that is, σ_a^2) multiplied by the autoregression coefficient φ_1. The general solution for the extrapolation of an AR (p) sequence at the lead time τ is

$$\hat{x}_t(\tau) = \varphi_1 \hat{x}_t(\tau - 1) + \cdots + \varphi_p \hat{x}_t(\tau - p) \tag{6.6}$$

where $\hat{x}_t(\tau - k) = x_{t+\tau-k}$ are the known time series elements if $\tau \leq k$.

Let

$$\varepsilon_t(\tau) = x_{t+\tau} - \hat{x}_t(\tau) \tag{6.7}$$

be the error of extrapolation from time t at lead time τ; its variance $\sigma_\varepsilon^2(\tau)$ can be defined in the following manner. In the operator form, Eq. (6.1) is

$$x_t = (1 - \varphi_1 B - \cdots - \varphi_p B^P)^{-1} a_t \tag{6.8}$$

or

$$x_t = \mathbf{\Phi}^{-1}(B) a_t. \tag{6.9}$$

Here $\mathbf{\Phi}^{-1}(B)$ is the operator inverse to the autoregressive operator in Eq. (6.1) and B is a backward shift operator, that is, $B^k x_t = x_{t-k}$. The extrapolation error can now be written as

$$\varepsilon_t(\tau) = a_{t+\tau} + \psi_1 a_{t+\tau-1} + \cdots + \psi_{\tau-1} a_{t-1}, \tag{6.10}$$

where ψ_j are coefficients of the operator $\mathbf{\Psi}(B) = \mathbf{\Phi}^{-1}(B)$. As the values of innovation sequence a_t are not correlated with each other, the variance of extrapolation error is found from Eq. (6.10) as

$$\sigma_\varepsilon^2(\tau) = (1 + \psi_1^2 + \cdots + \psi_{\tau-1}^2)\sigma_a^2. \tag{6.11}$$

Equations (6.6) and (6.11) define the extrapolation function and the variance of extrapolation error, respectively.

If the correct AR coefficients $\varphi_j, j = 1, \ldots p$ (and, therefore, the correct coefficients $\psi_j, j = 1, \ldots, \tau - 1$) are replaced with coefficients $\hat{\varphi}_j \neq \varphi_j$, the error variance will obviously exceed the value given with Eq. (6.11). This means that the approach to extrapolation within the Kolmogorov–Wiener theory ensures the least possible error variance of linear extrapolation. A fuller and stricter description of the theory can be found in Yaglom (1962, Chap. 4) and in Box et al. (2015, Chap. 5). It should be stressed that the theory created by A. Kolmogorov and N. Wiener is fundamental for the forecasting task because any linear method that does not agree with their theory produces worse results. If the time series is Gaussian, this statement is true for any method of forecasting.

If the time series' PDF is symmetric with respect to the distribution mode, the extrapolation will show the most probable future behavior of the process.

Summing up, the stages of linear least-squares extrapolation within the Kolmogorov–Wiener theory and with autoregressive modeling are

- fit a proper AR model (6.1) to the time series,
- calculate the forecasts according to Eq. (6.6), and
- determine the forecast error variance according to Eq. (6.11).

The confidence bounds for the forecasted values are calculated assuming a Gaussian distribution for the extrapolation error.

The efficiency of forecasting (and the degree of time series persistence) is measured here with the prediction quality criterion (PQC)

$$r_e(\tau) = \sqrt{1 - \sigma_\varepsilon^2(\tau)/\sigma_x^2} \tag{6.12}$$

and with the relative predictability criterion (RPC), which presents a measure of prediction error variance:

$$\rho(\tau) = \sigma_\varepsilon(\tau)/\sigma_x. \tag{6.13}$$

Obviously, $\rho(\tau) = \sqrt{1 - r_e^2(\tau)}$. The criterion (6.12) coincides with the correlation coefficient between the unknown future value $x_{t+\tau}$ and its forecast $\hat{x}_t(\tau)$ from time t at lead time τ. The relative predictability criterion (6.13) shows the degree of vicinity of the error's standard deviation to the standard deviation of the time series. Both criteria stay between 0 and 1 and present monotonic functions of the lead time τ. If the variance $\sigma_\varepsilon^2(1) = 0$, it will be zero at all lead times.

Another predictability criterion is the limit of statistical predictability τ_α; it can be defined as the lead time τ at which the distance between the confidence limits and the extrapolation trajectory reaches a specific value at a confidence level α. A reasonable choice for this quantity would be the lead time at which the confidence limits equal $\pm\sigma_x$ with respect to $\hat{x}_t(\tau)$. For the confidence level 0.9, the limit of predictability $\tau_{0.9}$ is reached when the root mean square value of the extrapolation variance $\rho(\tau)$ becomes equal to 0.6.

The forecasting method based upon the AR modeling of time series does not require the knowledge of the spectral density. But, it is strongly recommended that the spectrum estimate be obtained through a Fourier transform of the time domain AR model (6.3). Respective equation is given in Chap. 3.

Unfortunately, the Kolmogorov–Wiener theory of extrapolation seems to be unknown in Earth sciences. For example, in 2018–2019, all journals published by the American Meteorological Society printed well over a hundred articles involving statistical forecasting and none of them contains a referral to the Kolmogorov–Wiener theory. The statistics for the *International Journal of Climatology* is also zero. Monographs on time series analysis and forecasting published decades ago (e.g., Epstein 1985) or recently (e.g., Hyndman and Athanasopoulos 2018, Wei 2019) do not mention KWT. The same is true of the books by von Storch and Zwiers (1999), van den Dool (2007), Wilks (2011), Mudelsee (2014), and Maity (2018) written for climatologists and geophysicists. Seemingly, the only examples of using KWT for extrapolation of geophysical processes belong to this author (Privalsky 1992, 2014).

This omission is regrettable because it is the KWT solution of the statistical forecasting problem for stationary time series that has the smallest possible error variance, or the smallest possible error variance of linear extrapolation in the non-Gaussian case. (The books by Box et al (2015) and Shumway and Stoffer (2017) do recognize and discuss the Kolmogorov–Wiener theory of extrapolation.)

The above-described autoregressive method of extrapolation can be extended to the multivariate case with the scalar AR coefficients becoming matrices and Eq. (6.6) transformed into a system of equations containing contributions from other components of the time series. The coefficients in Eq. (6.11) and all variances also become matrices. A brief analysis of prediction quality for a bivariate meteorological time series (MJO) will be given in Chap. 8.

Aside from the tides, the geophysical processes that possess significant statistical predictability are rare. To some degree, they include climatic time series obtained by averaging the data over large parts of the planet. Otherwise, high predictability can happen due to the presence of strong quasi-periodic phenomena. The examples given here include the global annual surface temperature, the ENSO components NINO and SOI, the Quasi-Biennial Oscillation (QBO), and the Madden–Julian Oscillation (MJO).

It should be stressed that any linear statistical forecast of a stationary time series will be incorrect if the method of forecasting disagrees with the Kolmogorov–Wiener theory of extrapolation. This statement is equally true for the scalar and multi-variate time series. Any linear method of statistical extrapolation that deviates from KWT produces forecasts whose error variance exceeds the value following from the theory. Thus, judging by the typical shape of the climate spectrum, the statistical predictability of climate (that is, when the time scales are measured in years) is generally low. However, the low statistical predictability is a property of climate variability and not the fault of the methods lying within the framework of KWT. In other words, looking for other approaches to forecasting stationary time series requires an explanation of why the Kolmogorov–Wiener theory is inapplicable or where it errs. The extrapolation with stationary ARMA models agrees with KWT, but the theory is still not mentioned in older and recent publications on the subject of forecasting through ARMA modeling, for example, in Abraham and Ledolter (2005) and in Papacharalampous et al. (2018). The software for ARMA modeling of stationary time series within the KWT framework is available, for example, in Shumway and Stoffer (2017) and in MATLAB.

In the extrapolation examples given below, the mean values of the time series coincide with the values corresponding to the initial data and they may be different from zero.

6.3   Example 1. Global Annual Temperature

According to Table 5.2, the higher predictability occurs for the annual surface temperature averaged over very large areas, up to the entire surface of the planet. This happens because respective time series contain most of their energy within the low-frequency part of the spectrum.

The global annual temperature (notated here as GLOBE) from 1850 (#1 in Appendix below and Fig. 6.1a) shows two intervals with a definite positive trend; the trend is longer and slightly faster during the latest several decades. Similar to the earlier interval from 1911 through 1944, the trend that happened during the years from 1974 through 2010 (the initial year for our extrapolation test below) may have been caused by natural factors (Privalsky and Fortus 2011) so that its higher predictability could have been the result of regular variations of climate. As for the higher frequencies, the spectral density estimate for the detrended time series (Fig. 6.1b) proves

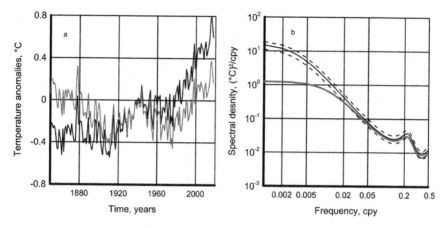

Fig. 6.1 Anomalies of annual global surface temperature, 1850–2018 (**a**) and their spectra (**b**). Detrended versions are shown in gray

that detrending the time series does not affect the spectrum at frequencies above 0.02 cpy (at time scales 50 years and shorter).

The goal of this test is to get an idea of extrapolation efficiency for the original and detrended time series. With year 2010 as the initial time for extrapolation, one has eight observed values of temperature anomalies that can be compared with predictions for 2011–2018.

The entire time series of GLOBE from 1850 through 2018 can be regarded as a sample of a stationary random process and extrapolated in accordance with its best fitting AR model. The second approach regards the time series as nonstationary: the sum of a stationary process plus trend (linear, in our case). The first version means that the trend is a part of the low-frequency variations caused by the natural climate variability; in the second version, the climate variability is regarded as stationary while the trend is caused by some external factors, including possible anthropogenic effects.

In both versions, the time series with the trend present or deleted behave as stationary autoregressive sequences of order $p = 4$. Judging by the shape of its spectral density (Fig. 6.1b), the original time series may be close to being nonstationary due to the dominant role of low-frequency variations but the roots of its characteristic equation significantly exceed 1 so that it should be treated as stationary. It also passes the test for stationarity recommended in Chap. 4 though the number of independent observations in the entire time series with the trend present and in its halves are small.

The predictability properties of the time series for both cases are shown in Fig. 6.2. In the first (stationary) case, the RPC increases to 0.6 at the lead time 5 years; that is, the time series with the trend regarded as a nature caused phenomenon has a high statistical predictability (Fig. 6.2a). In the second case (stationarity plus trend), the predictability limit $\tau_{0.7}$ is just one year (Fig. 6.2b).

The results of the GLOBE extrapolation test in accordance with the two versions of the time series are given in Fig. 6.3. The values of GLOBE observed from 2011

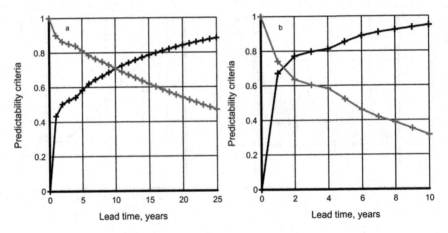

Fig. 6.2 Statistical predictability of the GLOBE time series, 1850–2018: PQC (black) and RPC (gray) as a stationary time series (**a**) and as a stationary time series plus linear trend (**b**)

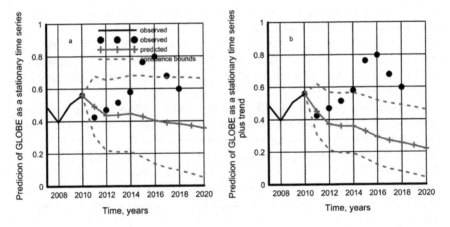

Fig. 6.3 Predicting GLOBE as a stationary time series (**a**) and as a stationary time series plus linear trend (**b**)

through 2018 and lying outside the 90% confidence limits (the dashed lines) mean unsatisfactory extrapolation. In the purely stationary case when the trend is not deleted (Fig. 6.3a), there are three or even two such failures. In the case of a sum of stationary time series with a linear trend, the extrapolation is less successful: five values of GLOBE observed after 2010 lie outside the confidence interval. Seemingly, the first approach produces better results. The transition from the stationary model to a mixture of a stationary model with trend results in two important changes: the extrapolation function tends to the mean value of the time series faster than in the stationary case and the time series variance diminishes from 0.070 (°C)^2 to 0.024 (°C)^2.

A probabilistic forecast of GLOBE for 2019–2025 is given in Fig. 6.4. The latest known value for 2019 lies within the confidence limits. Such forecasts of natural climate variations could be taken into account in climate projections.

The results given in this section demonstrate that the nature-caused variations of globally averaged annual temperature possess statistical predictability at lead times up to 5 years and the predictability is better for the original time series. This statement also holds for oceanic and terrestrial annual temperature averaged over the northern hemisphere. At the same time, a higher predictability does not necessarily mean that the linear trend is caused only by the nature.

6.4 Example 2. Quasi-Biennial Oscillation

The Quasi-Biennial Oscillation will be discussed here at $\Delta t = 1$ month. The time series used for this example is QBO at the atmospheric pressure level 20 hPa, which corresponds to the altitude of about 26 km above mean sea level (Fig. 6.5a and #2 in Appendix). The spectral density estimate is shown in Fig. 6.5b with the frequency axis given in a linear scale.

At the time when this text was being written, monthly observations of QBO were available from January 1953 through April of 2019. The test extrapolation for the entire 2018 and the next six months of 2019, from May through November, which have been added in December 2019, is based upon the part of the time series that ends in December 2017 ($N = 780$).

The optimal, according to three of the five order selection criteria used here, is the AR model of order $p = 10$:

$$x_t = \varphi_1 x_{t-1} + \varphi_2 x_{t-2} + \cdots + \varphi_{10} x_{t-10} + a_t. \tag{6.14}$$

It means that the extrapolation equation is

$$\hat{x}_t(\tau) = \varphi_1 \hat{x}_t(\tau - 1) + \varphi_2 \hat{x}_t(\tau - 2) + \cdots + \varphi_{10} \hat{x}_t(\tau - 10). \tag{6.15}$$

The white noise variance corresponding to the AR(10) model is $\sigma_a^2 \approx 21$ (m/s)^2 while the total variance of wind speed at 20 hPa is $\sigma_x^2 \approx 389$ (m/s)^2. Therefore, the predictability criterion $\rho(1) \approx 0.05$ and the correlation coefficient (6.13) between the unknown true and predicted values of wind speed at lead time $\tau = 1$ month is 0.97. As seen from Fig. 6.6, the statistical predictability of QBO at the 20 hPa level described with the predictability criteria $r_e(\tau)$ and $\rho(\tau)$ is quite high.

The results of prediction test with the initial time in December 2017 (Fig. 6.7a) show that the AR method of extrapolation is working quite well with this time series: 19 of the 20 monthly forecasts stay within the 90% confidence limits. More predictions are given from December 2018 through January 2021 for future verification (Fig. 6.7b). The data used for the AR models were from January 1953 through December 2017 and through December 2018, respectively. By the time when the

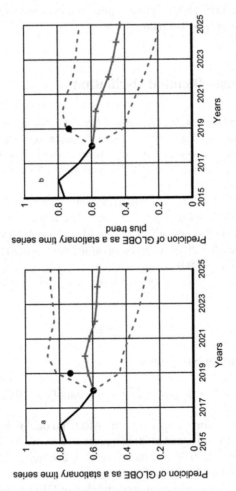

Fig. 6.4 Probabilistic forecast of GLOBE for 2019–2025 as a stationary time series (**a**) and as a stationary time series plus linear trend (**b**). The black circles show the annual temperature observed in 2019

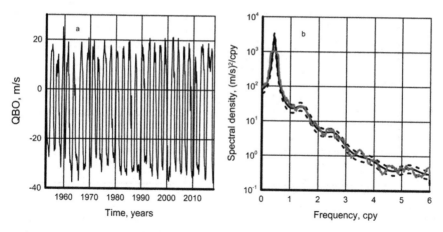

Fig. 6.5 Monthly wind speed at the 20 hPa level, January 1953–January 2018 (**a**) and its AR (black) and MTM (gray) spectrum estimates (**b**)

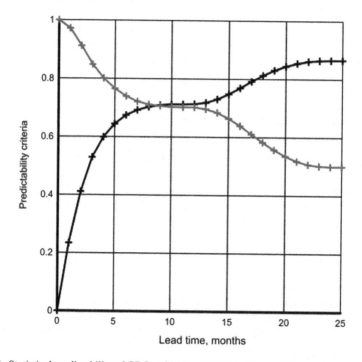

Fig. 6.6 Statistical predictability of QBO at 20 hPa: PQC (black) and RPC (gray)

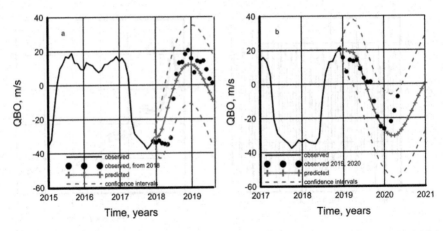

Fig. 6.7 Observed (black) and predicted (gray) wind speed at the 20 hPa level of QBO with initial prediction dates December 2017 (**a**) and December 2018 (**b**)

book was ready for the publisher, more observations became available and they are included into Fig. 6.7b. The quality of extrapolation seems to be high, but one should have in mind that the 90% confidence intervals shown in the figure are wide.

6.5   Example 3. ENSO Components

Predicting the behavior of the oceanic ENSO component—sea surface temperature in equatorial Pacific—is regarded as a very important task in climatology and oceanography (e.g., #3 and #4 in Appendix). Attempts to predict ENSO's atmospheric component—the Southern Oscillation Index—do not seem to be numerous (e.g., Kepenne and Ghil 1992). In this section, both tasks will be treated within the KWT framework.

At the annual sampling rate, the ENSO components behave similar to white noise (Chap. 5); their predictions through any probabilistic method would be practically useless. In the current example, the statistical forecasts of sea surface temperature in the ENSO area NINO3 (5 °N–5 °S, 150 °W–90 °W) and the Southern Oscillation Index are executed at a monthly sampling rate using the data from January 1854 through February 2019 and from 1876 through February 2019, respectively. The data are available at websites #5 and #6 given in Appendix below. The NINO3 time series is shown in Fig. 6.8a. It can be treated as a sample of a stationary process.

The autoregressive analysis of this time series showed an AR(5) model as optimal. Its spectral density estimates are shown in Fig. 6.8b. The low-frequency part of the spectrum up to 0.5 cpy contains about 70% of the NINO3 variance and the ratio of the white noise RMS to the NINO3 RMS is 0.39. In contrast to the annual global

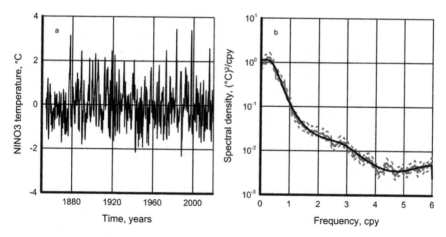

Fig. 6.8 Time series NINO3 (**a**) and its AR (black) and MTM (gray) spectra (**b**)

temperature with the trend present, the predictability of NINO3 diminishes quite fast, but, as seen from Fig. 6.9, it still extends to several months.

A KWT prediction from the end of 2017 through January 2019 is given in Fig. 6.10a. The result of the test turned out to be satisfactory but one has to remember that the 90% confidence limits for the extrapolated values are rather wide. Only the first four or five predicted values lie within the relatively narrow interval not exceeding $\pm\sigma_x$.

Fig. 6.9 Statistical predictability of NINO3: PQC (black) and RPC (gray). The dashed line is the prediction quality criterion for SOI

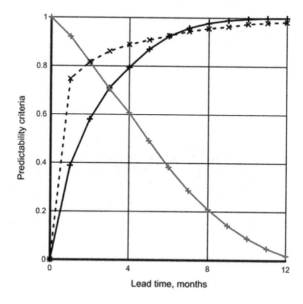

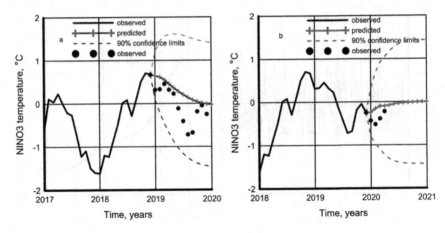

Fig. 6.10 Observed (black) and predicted (gray) NINO3 from December 2018 (**a**) and from December 2019 (**b**)

The KWT prediction of NINO3 has been executed in March 2019 when only two values of NINO3 for 2019 were known from observations. As seen from Fig. 6.10a, the values observed later through November 2019 lie within the confidence limits. The data for the first four months of 2020 have been added in May 2020. The initial point for that forecast was February 2019.

The probability density function of the monthly NINO3 time series is not Gaussian; a good approximation for its PDF is a three-parameter log-logistic distribution, and it is shaped in such a way that the nonlinear approach would hardly improve the KWT results in a significant manner.

The atmospheric component of ENSO is the Southern Oscillation Index (Fig. 6.11a). The optimal model for this time series at $\Delta t = 1$ month is AR(14); its

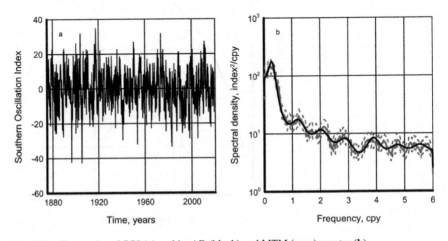

Fig. 6.11 Time series of SOI (**a**) and its AR (black) and MTM (gray) spectra (**b**)

spectral estimates are shown in Fig. 6.11b. The dynamic range of the spectrum in this case is an order of magnitude smaller than in the NINO3 spectrum (Fig. 6.8b) meaning that the atmospheric component of ENSO is closer to a white noise than the oceanic component NINO3. According to both the AR (MEM) and MTM estimates, the spectrum contains a statistically significant peak at $f \approx 0.22$ cpy, but the role of the peak is not critical because the ratio of extrapolation root mean square error to the SOI's root mean square value is close to 0.75 even at the lead time $\tau = 1$ month. The contribution of NINO3 spectrum to its variance at frequencies between zero and 0.5 cpy is about 70% while in SOI it amounts to 55%. The predictability criterion PQC for the SOI time series shown in Fig. 6.9 becomes close to one faster than in the NINO3 case, which leads to a wider confidence interval for predicted values. Due to these features, the statistical forecasting of SOI within KWT would not be of any practical value. Though the time series is not Gaussian, a nonlinear approach to its extrapolation would hardly improve the results of forecasting in a significant manner just because its PDF is close to symmetrical.

6.6   Example 4. Madden–Julian Oscillation

This data set is taken from site #7 in Appendix below. As mentioned in Chap. 5, the components of MJO regarded as samples of scalar processes may possess relatively high statistical predictability. At the unit lead time (one day), the statistical predictability criterion $\rho(1)$ for the RMM1 component equals to about 0.18 and the process should be studied in more detail. The predictability of the RMM1 time series decreases rather fast (Fig. 6.12), but it stays acceptable up to 6 days. The RMM2 component behaves in the same way.

Prediction examples (Fig. 6.13) turned out to be rather successful even for longer lead times, but the confidence bounds are rather wide. The cycles with periods close to 50 days cannot be reliably reproduced by the extrapolation trajectory at lead times close to the period of the cycle.

If the sampling rate is increased from 1 day to 10 days, the resulting time series becomes poorly predictable even at the unit lead time, that is, at 10 days. As both the original time series RMM1 and RMM2 and the time series with $\Delta t = 10$ days are Gaussian or close to Gaussian, one can say that the Madden–Julian Oscillation is practically unpredictable at that sampling rate in spite of the presence of a significant spectral maximum.

The examples in this chapter include five rather typical and at the same time dissimilar cases with the sampling rates of one year, one month, and one day; they can be summed up in the following way:

- the global surface temperature that has some predictability due to the dominant role of low-frequency variations even when the linear trend is deleted.
- highly predictable Quasi-Biennial Oscillation whose spectrum contains a powerful peak at the low-frequency part of the spectrum,

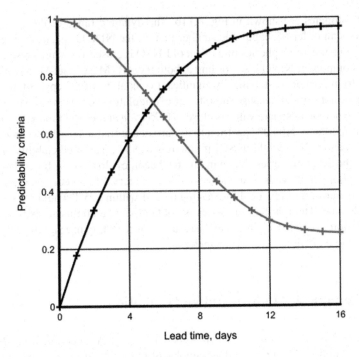

Fig. 6.12 Statistical predictability of MJO component RMM1

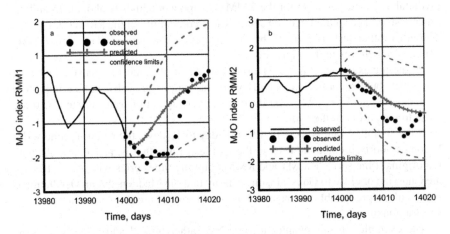

Fig. 6.13 Forecasting MJO components RMM1 (**a**) and RMM2 (**b**)

- NINO3—the oceanic component of ENSO—with a relatively dominant low-frequency variability, satisfactory forecasts, and quickly increasing error variance,

- SOI—the atmospheric component of ENSO—which contains a statistically significant spectral maximum and still has low predictability because of the low dynamic range of its spectrum.
- MJO, with its smooth spectral maximum and acceptable forecasts at several lead times.

In conclusion, it has been shown here that the use of a forecasting method which agrees with the Kolmogorov–Wiener theory of extrapolation produces satisfactory results if the spectrum of the time series is concentrated within a relatively narrow frequency band. If the spectrum is spread more or less evenly over frequency, the time series is practically unpredictable. In all cases, even when the latest and previously unknown values of the time series lie close to the predicted trajectory, one should keep in mind the width of the confidence interval as a function of lead time. It is the quantity that defines the usefulness of forecasts.

Appendix

1. https://crudata.uea.ac.uk/cru/data/temperature/HadCRUT4-gl.dat
2. https://www.geo.fu-berlin.de/met/ag/strat/produkte/qbo/qbo.dat
3. https://www.esrl.noaa.gov/psd/enso/enso.forecast.html
4. https://iri.columbia.edu/our-expertise/climate/forecasts/enso/current/
5. https://climexp.knmi.nl/data/iersst_nino3a_rel.dat
6. http://www.bom.gov.au/climate/current/soi2.shtml
7. http://www.bom.gov.au/climate/mjo/graphics/rmm.74toRealtime.txt.

References

Abraham B, Ledolter J (2005) Statistical methods for forecasting. Wiley-Interscience, Hoboken

Box GEP, Jenkins GM (1970) Time series analysis. Forecasting and control. Wiley, Hoboken

Box G, Jenkins M, Reinsel G, Liung G (2015) Time series analysis. Forecasting and control, 5th edn. Wiley, Hoboken

Coales J, Kahne S (2014) "The Yellow Peril" and after. IEEE Control Systems Magazine, February 2014

Epstein ES (1985) Statistical inference and prediction in climatology, a Bayesian approach. American Meteorological Society, Boston

Fortus M (1978) An explicit formula for the mean square error of a linear extrapolation of a stationary random process with rational spectral density. Dokl Akad Nauk SSSR 241:34–36

Hyndman R, Athanasopoulos G (2018) Forecasting: principles and practice. OTexts, Melbourne, Australia

IPCC (2013) Climate Change 2013: The Physical Science Basis. In: Stocker TF, Qin D, Plattner G-K, Tignor M, Allen SK, Boschung J, Nauels A, Xia Y, Bex V, Midgley PM (eds) Contribution of Working Group I to the fifth assessment report of the intergovernmental panel on climate change. Cambridge University Press, Cambridge, United Kingdom and New York, NY, USA, 1535 pp

Kepenne C, Ghil M (1992) Adaptive filtering and prediction of the Southern Oscillation Index. J Geophys Res 97:20,449–20,454

Kolmogorov AN (1939) Sur l'interpolation et extrapolation des suites stationnaires. Compte Randu de l'Academie des Sciences. Paris 208:2043–2046

Kolmogorov AN (1941) Interpolation and extrapolation of stationary random sequences. Compte Randu de l'Academie des Sciences de l'URSS. Ser Math 5:3–14

Lorenz E (1963) The predictability of hydrodynamic flow. T New York Acad Sci Ser II 25:409–432

Lorenz E (1975) Climatic predictability. In: The physical basis of climate and climate modelling. WMO, Geneva, Appendix 2.1, pp 132–136

Lorenz E (1995) Predictability: a problem partly solved. In: Proceedings of Seminar on predictability, vol 1, pp 1–18, ECMWF, Reading, UK

Lorenz E (2007) Foreword. In: van den Dool H (ed) Empirical methods in short-term climate predictions. NOAA Climate Prediction Center. Oxford University Press, Oxford. pp XI–XII

Maity R (2018) Statistical methods in hydrology and hydroclimatology. Springer Nature, Singapore

Mudelsee M (2014) Climate time series analysis, 2nd edn. Springer, Heidelberg

Munk W, Cartwright D (1966) Tidal spectroscopy and prediction. Philos Trans R Soc 259(A1105):533–581

Papacharalampous G, Tyralis H, Koutsoyiannis D (2018) One-step ahead forecasting of geophysical processes within a purely statistical framework. Geoscience Lett 5:12

Privalsky V (1992) Statistical analysis and predictability of Lake Erie water level variations. J Great Lakes Res 18:236–243

Privalsky V (2014) Statistical prediction of global surface temperature for use with GCM-based projections of climate. Proc ITISE 2014 Granada

Privalsky V, Fortus M (2011) On possible causes of global warming. Theor Probab Appl 56:313–317. https://doi.org/10.1137/S0040585X97985418

Shumway R, Stoffer D (2017) Time series analysis and its applications. Springer, Berlin

van den Dool H (2007) Empirical methods in short-term climate predictions. NOAA Climate Prediction Center. Oxford University Press, Oxford, pp 11–12

von Storch H, Zwiers F (1999) Statistical analysis in climate research. Cambridge University Press, Cambridge

Wei W (2019) Multivariate time series analysis and applications. Wiley, Hoboken

Wiener N (1949) Interpolation, extrapolation and smoothing of stationary time series. Wiley, New York

Wilks D (2011) Statistical methods in the atmospheric sciences, 3rd edn. Elsevier, Amsterdam

Yaglom AM (1962) An introduction to the theory of stationary random functions. Prentice-Hall International, London

Zade L, Ragazzini J (1950) An extension of Wiener's theory of prediction. J Appl Phys 21:645–656

Zasukhin V (1941) On theory of multidimensional random processes. Dokl Acad Nauk SSSR 33:435–437 (in Russian)

Chapter 7
Bivariate Time Series Analysis

Abstract The autoregressive analysis of stationary multivariate time series includes construction of a time domain model in the form of a multivariate stochastic difference equation and its transformation into a spectral matrix to study frequency domain properties. The time series is treated as a linear system with one output process and one or several inputs. This approach provides an explicit time domain description of the time series as a function of the past of output and input components. In the bivariate case, the frequency domain results include spectral densities, coherent spectrum, coherence function, and gain and phase factors. The coherence function shows the degree of linear interdependence, the coherent spectrum presents the contribution of the input component to the output, the gain factor consists of amplification coefficients, and the phase factor describes lags between the time series. All these quantities are frequency dependent. The Granger causality concept and feedbacks within the system are discussed as a part of bivariate time and frequency domain analysis. Some comments are given regarding the software for parametric analysis of multivariate time series. The use of cross-correlation coefficient and regression equation for describing dependence between time series or for reconstructing them is shown to be incorrect.

Multivariate data analysis in Earth and solar sciences usually involves two tasks: studying interdependence between time series (what is called teleconnections in climatology) and reconstructing time series. The goal of the first task is to find a geophysical or solar process represented with a time series which would be statistically related to another time series at the same or different location. A common example is looking for a dependence between an ENSO component and some other climate variable such as surface temperature, precipitation, Palmer Drought Severity Index, etc. (e.g., Yang et al. 2019). Other applications may include a search for correlations between different atmospheric variables (e.g., Bhowmik and Sankarasubramanian 2019), the response of mountain ice volume to indices of atmospheric circulation (Salinger et al. 2019) or looking for a Sun-induced peak in the spectrum of East Asian summer monsoon (Jin et al. 2019).

The second major task is to determine the unknown past values of a geophysical or solar time series (called the target) through its statistically evaluated relation to

© Springer Nature Switzerland AG 2021
V. Privalsky, *Time Series Analysis in Climatology and Related Sciences*,
Progress in Geophysics, https://doi.org/10.1007/978-3-030-58055-1_7

another time series (called the proxy). The proxy itself can be a multidimensional time series (e.g., Kumar et al. 2018). The proxy is represented with time series covering the entire interval of observations of the target process and extending into the past. The unknown past values of the shorter target time series can be calculated (reconstructed, restored) by using the dependence between the target and proxy time series obtained for the interval of their simultaneous observations. In solar sciences, it can be a reconstruction of total or spectral solar irradiance (the target) using the time series of solar sunspots as a proxy. In Earth sciences, it can be a restoration of unknown previous parts of time series of surface temperature, precipitation, river flow, climate index, or other characteristics with the help of proxy time series such as tree ring widths, thickness of deposition layers, and concentration of isotopes in ice kerns. Recent examples can be found in Schweiger et al. (2019), Simpson et al. (2019), Tan et al. (2019), Zhang et al. (2019), Khan et al. (2020).

Other tasks that require multivariate analysis may exist in the Earth and solar sciences but searching for teleconnections and reconstructing time series seem to play the domineering role. Hundreds of publications have been dedicated to these subjects since decades in the past; new research is being printed on a regular basis in climatology and paleoclimatology, in solar physics, hydrology, and in practically all other Earth sciences.

The traditional solutions for the tasks of teleconnection research and time series reconstruction have a common basis: the degree of success in solving the task is based upon the value of the ordinary or multiple cross-correlation coefficient between the time series that are being analyzed. In some cases, the maximum absolute value of cross-correlation function estimate is selected instead of the cross-correlation coefficient; this latter technique does not change anything because relations between time series in the time domain is characterized with the cross-correlation function so that a single value of the function is useless for that purpose.

The mathematical tool that is practically always applied for time series reconstruction is the linear regression equation (Christiansen and Ljungkvist 2017). However, both the cross-correlation coefficient and the regression equation belong to the area of classical mathematical statistics while the time series analysis is based upon theory of random processes and information theory. Besides, the regression equation is unacceptable because of physical considerations. Firstly, physically realizable systems always contain a delay, which normally does not exist in the regression equation. Secondly, the regression equation means that the spectrum of the output process has the same shape as the spectrum of the input, which cannot be true for all target/proxy systems. The cross-correlation coefficient and the regression equation are designed for analysis of time-invariant random vectors and cannot be used for analysis of linear dependences between random functions of time, that is, between time series.

An example of inability of the cross-correlation coefficient to measure relations between time series had been given quite some time ago in Privalsky and Jensen (1995a). If $x_{1,t}$ is a white noise time series and $x_{2,t} = x_{1,t+1}$, the cross-correlation coefficient between $x_{1,t}$ and $x_{2,t}$ is zero though the time series are strictly linearly related to each other. This example is sufficient by itself but with the cross-correlation coefficient being so deeply engrained in many Earth and solar sciences, it should be discussed in more detail.

Let $x_{1,t}$, $t = 1, \ldots, T$, where $T = N\Delta t$, be a scalar time series generated by a stationary random process different from white noise. Obviously, the cross-correlation coefficient of the time series x_t with itself is unity. Let the second time series be $x_{2,t} = x_{1,t+1}$, that is, $x_{1,t}$ is now shifted by one time unit forward, so that the lengths of the time series to be compared with each other becomes $(N-1)\Delta t$. The value of the cross-correlation coefficient will diminish and cannot be equal to unity anymore. Again, as the time shift is a linear operator, the time series $x_{1,t}$ and its shifted version $x_{2,t}$ are linearly related to each other but the dependence between them, if characterized with the cross-correlation coefficient, ceased to be strictly linear. This means that the correlation coefficient does not reflect the degree of linear dependence between time series. The following example of the same inability is even more general.

Let $x_t, t = 1, \ldots, T$, be a zero mean scalar time series generated by any stationary autoregressive (AR), moving average (MA), or mixed autoregressive and moving average (ARMA) random process, that is, let x_t, be an ARMA (p, q) time series. This means that it is described with the following stochastic difference equation:

$$x_t - \varphi_1 x_{t-1} - \cdots - \varphi_p x_{t-p} = a_t - \theta_1 a_{t-1} - \cdots - \theta_1 a_{t-q} \tag{7.1}$$

or

$$x_t = (1 - \varphi_1 B - \cdots - \varphi_p B^p)^{-1}(1 - \theta_1 B - \cdots - \theta_q B^q)a_t, \tag{7.2}$$

where a_t is white noise and B is a backshift operator: $B^j x_t = x_{t-j}$. As the product of linear operators in the right-hand part of the last equation is also a linear operator, the time series x_t presents the result of a linear transformation of the innovation sequence a_t. In spite of this, the cross-correlation coefficient between x_t and a_t can be equal to unity if and only if $x_t = a_t$. In particular, if $x_t = \varphi_1 x_{t-1} + a_t$ with $|\varphi_1| < 1$, that is, if x_t is a stationary Markov chain, its cross-correlation coefficient with a_t is $\sqrt{(1 - \varphi_1^2)}$, which is not a unity for any $\varphi_1 \neq 0$. The transformation of a_t into x_t is linear but the cross-correlation coefficient between x_t and a_t is not unity.

The conclusion from these examples is clear:

- the cross-correlation coefficient or any other single value of the cross-correlation function cannot serve as a measure of linear dependence between time series.

The following statement about the regression equation is also true:

- the regression coefficient cannot serve as a quantitative measure of linear dependence between time series; therefore, the regression equation is not applicable to time series.

Even if the cross-correlation coefficient or any other element of the cross-correlation function seems to be very high, it does not mean that it is statistically significant. The reason for this is that the error variances of time domain statistics of multivariate time series depend upon the behavior of the entire covariance and cross-covariance functions (e.g., Yaglom 1987; Bendat and Piersol 2010; also see Chap. 4).

The above statements are also true for multivariate time series consisting of more than two scalar components. The inapplicability of the cross-correlation, regression coefficient, and regression equation in time series analysis follows from the fact that time series present random functions of time, while random vectors do not depend upon time. The examples given above and their discussion lead to an inevitable conclusion that time series must be treated as random functions of time, not as time-invariant sets of random variables. This property of random function has been known in information theory since 1957 (see below).

7.1 Elements of Bivariate Time Series Analysis

A multivariate random process is a set of scalar random processes:

$$\mathbf{x}_t = [x_{1,t}, \ldots, x_{M,t}]', \tag{7.3}$$

where t is time, M the dimension of the process (the number of scalar components in $\mathbf{x}_t$), and the strike means matrix transposition. The components of $\mathbf{x}_t$ are characterized with respective scalar and joint PDFs. If all scalar PDFs are Gaussian, the joint PDFs are also Gaussian and the process $\mathbf{x}_t$ is Gaussian as well (e.g., Yaglom 1987). The properties of stationarity and ergodicity have the same meaning as in the case of scalar processes. In this part of the book, the time series will be regarded as Gaussian, which often agrees with climate observations (see Privalsky and Yushkov 2018, and Chap. 5) and with many other geophysical processes, especially those that do not contain quasi-periodic components such as daily and seasonal trends. If the process is not Gaussian, its analysis including the second statistical moments (covariance and spectral matrices) remains the same as in the Gaussian case.

The task of analyzing a sample record of a multivariate random process means obtaining the same statistical characteristics as for a scalar process plus the moments of the joint PDF. The quantities that need to be obtained include the multivariate stochastic difference equation, that is, an autoregressive (in our case) model in the time domain, and the spectral matrix in the frequency domain. This and the following chapters up to Chap. 14 are dedicated to the bivariate case ($M = 2$).

7.1.1 Bivariate Autoregressive Models in Time Domain

The AR model of a bivariate zero mean time series $\mathbf{x}_t = [x_{1,t}, x_{2,t}]'$ is

$$x_{1,t} = \varphi_{11}^{(1)} x_{1,t-1} + \varphi_{12}^{(1)} x_{2,t-1} + \cdots + \varphi_{11}^{(p)} x_{1,t-p} + \varphi_{12}^{(p)} x_{2,t-p} + a_{1,t}$$

$$x_{2,t} = \varphi_{21}^{(1)} x_{1,t-1} + \varphi_{22}^{(1)} x_{2,t-1} + \cdots + \varphi_{21}^{(p)} x_{1,t-p} + \varphi_{22}^{(p)} x_{2,t-p} + a_{2,t} \qquad (7.4)$$

or

$$\mathbf{x}_t = \sum_{j=1}^{p} \mathbf{\Phi}_j \mathbf{x}_{t-j} + \mathbf{a}_t. \qquad (7.5)$$

Here, p is the AR order,

$$\mathbf{\Phi}_j = \begin{bmatrix} \varphi_{11}^{(j)} & \varphi_{12}^{(j)} \\ \varphi_{21}^{(j)} & \varphi_{22}^{(j)} \end{bmatrix} \qquad (7.6)$$

are the matrix AR coefficients and $\mathbf{a}_t = [a_{1,t}, a_{2,t}]'$ is a bivariate white nose innovation sequence with a covariance matrix

$$\mathbf{P_a} = \begin{bmatrix} P_{11} & P_{12} \\ P_{21} & P_{22} \end{bmatrix}, \qquad (7.7)$$

where P_{11}, P_{22} are the variances of $a_{1,t}, a_{2,t}$ and $P_{12} = P_{21} = \mathrm{cov}[a_{1,t}, a_{2,t}]$ is their covariance. The notation for multivariate autoregressive model of order p is **AR** (p).

Equations (7.4) and (7.5) represent a linear stochastic system, which possesses physically reasonable features:

- a time delay between the current values and system's past behavior.
- each process may depend upon its own past.
- each process may depend upon the other process's past.

The time series $x_{1,t}$ and $x_{2,t}$ will be regarded in what follows as the output and input of the linear system described with these equations. The frequency domain analysis and physical considerations allow one to verify in most cases whether this decision is correct. If it turns out that $x_{1,t}$ and $x_{2,t}$ are the input and output, the components of the time series can be interchanged.

The bivariate autoregressive model described with Eqs. (7.4) or (7.5) provides valuable information about properties of the process $\mathbf{x}_t$. In particular, it shows

- the memory of the process $\mathbf{x}_t$, which is defined with the autoregressive order p—the number of past values of $\mathbf{x}_t$ that should be taken into account,
- in what way the time series $x_{1,t}$ depends upon its past values $x_{1,t-1}, \ldots, x_{1,t-p}$: the coefficients $\varphi_{11}^{(j)}$, $j = 1, \ldots, p$, provide quantitative measures of dependence between the current and past values of time series through their absolute values and their signs; the same is true for the time series $x_{2,t}$ and coefficients $\varphi_{22}^{(j)}$,
- in what way the component $x_{1,t}$ depends upon the past values of the time series $x_{2,t-j}$, $j = 1, \ldots, p$; the coefficients $\varphi_{12}^{(j)}$, $j = 1, \ldots, p$, provide quantitative

measures of dependence between the current value of $x_{1,t}$ and past values $x_{2,t-j}$; similar information about the dependence of $x_{2,t}$ upon $x_{1,t-j}$ is available through the coefficients $\varphi_{21}^{(j)}$ for the second component of the time series,

- the role played by the innovation sequence $\mathbf{a}_t = [a_{1,t}, a_{2,t}]'$; its components $a_{1,t}, a_{2,t}$ do not depend upon their past and constitute the unpredictable part of the bivariate process; therefore, the ratios P_{11}/σ_1^2 and P_{22}/σ_2^2 define the statistical predictability (or persistence) of the process' components $x_{1,t}$ and $x_{2,t}$ at the unit lead time: if the ratio is close to one, the time series predictability and persistence are low,
- the dependence between the innovation sequence components $a_{1,t}$ and $a_{2,t}$ expressed as the cross-correlation coefficient $\rho_{12} = P_{12}/\sqrt{P_{11}P_{22}}$.

If the time series $x_{1,t}$ depends upon past values of $x_{2,t}$ and the time series $x_{2,t}$ depends upon past values of $x_{1,t}$, Eq. (7.4) describes a linear feedback system, or a linear system with a closed feedback loop.

As the matrix (7.7) may not be diagonal, the time series $x_{1,t}$ and $x_{2,t}$ can be interdependent even if all AR coefficients are zero. Thus, the bivariate white noise

$$x_{1,t} = a_{1,t}$$
$$x_{2,t} = a_{2,t} \tag{7.8}$$

seems to be the only case when the regression equation becomes applicable to time series.

The values P_{11} and P_{22} are also used in analysis of feedback and causality—the concept introduced by C. Granger (1963) and discussed in a series of his later publications (see Sect. 7.2).

The statistics that describes the dependence between the components of bivariate time series in the time domain is the matrix covariance function

$$\mathbf{R}_{12}(k) = \begin{bmatrix} R_{11}(k) & R_{12}(k) \\ R_{21}(k) & R_{22}(k) \end{bmatrix}, \quad k = 0, \pm 1, \pm 2, \ldots \tag{7.9}$$

where the estimates of the covariance functions are obtained as

$$R_{ij}(k) = \frac{1}{N-k} \sum_{k=0}^{N-k} x_{i,t} x_{j,t+k} \tag{7.10}$$

and $R_{21}(k) = R_{12}(-k)$. The diagonal elements of the matrix (7.9) are the covariance functions, and the off-diagonal ones are the cross-covariance functions.

When the elements of the matrix (7.9) are divided by respective root mean square values, that is, $r_{ij}(k) = R_{ij}(k)/\sqrt{R_{ii}(0)R_{jj}(0)}$, the result is the correlation matrix

$$\mathbf{r}_{12}(k) = \begin{bmatrix} r_{11}(k) & r_{12}(k) \\ r_{21}(k) & r_{22}(k) \end{bmatrix}. \tag{7.11}$$

The diagonal elements of the matrix (7.11) are the correlation functions, and the off-diagonal elements are the cross-correlation functions.

As mentioned just above and in Chap. 3, statistical estimation of covariance or correlation functions is cumbersome because the estimation error variances for each individual function depend upon the behavior of the entire covariance (or correlation) matrix. The dependence between the components of bivariate time series is better described with frequency-dependent functions (the next subsection).

The software for analyzing multivariate AR models, including selection of the order p with order selection criteria (OSC), is available in MATLAB and in R (e.g., Shumway and Stoffer 2017) and we will just stress the necessity to follow order selection criteria in the autoregressive analysis of multivariate time series. An arbitrarily selected AR order is bound to lead to erroneous results. There may be some cases when the model's order is already known from physical consideration but, generally, it is imperative to use proper mathematical tools for choosing an order; the order selection criteria are probably the best for this purpose. Such criteria present multivariate versions of the OSCs used for the scalar time series (see Chap. 3).

Some order selection criteria that should be used for determining an optimal order for bivariate (and multivariate) time series are given in Reinsel (2003):

Akaike criterion $AIC_r \approx \log(|\hat{\mathbf{P}}_a|) + 2r/N$,

Schwarz criterion $BIC_r \approx \log(|\hat{\mathbf{P}}_a|) + r \log(N)/N$, and

Hannan–Quinn criterion $HQC_r \approx \log(|\hat{\mathbf{P}}_a|) + 2r \log(\log(N))/N$,

where $r = M^2 p$ and $|\hat{\mathbf{P}}_a|$ is the determinant of estimated covariance matrix of innovation sequence.

For teleconnection and reconstruct research, one needs to know contributions of scalar components of a bivariate time series to each other. A time domain solution is suggested below.

To determine the contribution of time series $x_{2,t}$ to $x_{1,t}$, multiply the first equation of (7.4) by $x_{1,t}$ and find the mathematical expectation. The result will be

$$\sigma_1^2 = \sum_{k=1}^{p} [\phi_{11}^{(j)} R_{11}(k) + \phi_{12}^{(j)} R_{21}(k)] + \rho_{12}^2 P_{11} + (1 - \rho_{12}^2) P_{11}, \qquad (7.12)$$

where $R_{11}(k)$ and $R_{21}(k)$ are the elements of the covariance matrix (7.9) and ρ_{12} is the correlation coefficient between the innovation sequences $a_{1,t}$ and $a_{2,t}$. The right hand part of Eq. (7.12) shows the structure of the variance σ_1^2 of the time series $x_{1,t}$. The terms $\phi_{11}^{(j)} R_{11}(k)$ and $\phi_{12}^{(j)} R_{21}(k)$ are the direct contributions to the variance of $x_{1,t}$ from its past values and from the past values of $x_{2,t}$, respectively. The term $\rho_{12}^2 P_{11}$ defines the share of the variance P_{11} of the innovation sequence $a_{1,t}$ linearly related to $a_{2,t}$, and the term $(1 - \rho_{12}^2) P_{11}$ shows the contribution of the part of $a_{1,t}$ not correlated to $a_{2,t}$.

Multiplying the second equation of (7.4) by $x_{2,t}$ and finding the mathematical expectation leads to equation

$$\sigma_2^2 = \sum_{k=1}^{p} [\phi_{21}^{(j)} R_{21}(k) + \phi_{22}^{(j)} R_{22}(k)] + \rho_{12}^2 P_{22} + (1 - \rho_{12}^2) P_{22}, \qquad (7.13)$$

that is, the structure of the variance σ_2^2 of the input time series is described in a similar way but in this case the first and second terms describe the contributions of $x_{1,t}$ and $x_{2,t}$ to σ_2^2. The term $\rho_{12}^2 P_{22}$ defines the share of the variance P_{22} of the innovation sequence $a_{2,t}$ linearly related to $a_{1,t}$ and the term $(1 - \rho_{12}^2) P_{22}$ shows the contribution of the part of $a_{2,t}$ not correlated with $a_{1,t}$. These two equations provide useful quantitative information about teleconnections. If the bivariate AR model contains a closed feedback loop, the equations serve as a time domain device to determine the impact of the input upon the output and vice versa.

7.1.2 Bivariate Autoregressive Models in Frequency Domain

The dependence of time series upon time also means that its behavior, including relations between its scalar components, may vary as a function of frequency. This is the reason why studying the frequency domain behavior of any time series, scalar or multivariate, is vital. The frequency-dependent properties of multivariate time series are determined through the spectral matrix $\mathbf{s}(f)$ of the time series $\mathbf{x}_t$. All frequency-dependent quantities can be estimated directly from the time series using nonparametric methods of spectral analysis but when an autoregressive model is used for the time domain it is more appropriate to estimate all required spectral characteristic through this time domain model. Besides, the nonparametric approach would lead to the loss of time domain information.

Equation (7.5) can be rewritten as

$$\mathbf{x}_t = (\mathbf{I} - \mathbf{\Phi}_1 B - \cdots - \mathbf{\Phi}_p B^p)^{-1} \mathbf{a}_t, \qquad (7.14)$$

where $\mathbf{I}$ is a $(p \times p)$ identity matrix. By doing a Fourier transform of the last equation (i.e., by changing the backshift operator B to $e^{-i2\pi f \Delta t}$) and finding the square of the modulus of both sides of the equation, one receives the spectral matrix of the time series $\mathbf{x}_t$ as

$$\mathbf{s}(f) = \frac{2\mathbf{P}_a \Delta t}{\left| \mathbf{I} - \sum_{j=1}^{p} \mathbf{\Phi}_j e^{-i2\pi jf \Delta t} \right|^2}, \quad 0 \le f \le 1/2\Delta t. \qquad (7.15)$$

The elements of the matrix

$$\mathbf{s}(f) = \begin{bmatrix} s_{11}(f) \, s_{12}(f) \\ s_{21}(f) \, s_{22}(f) \end{bmatrix} \tag{7.16}$$

are the spectral densities $s_{11}(f)$, $s_{22}(f)$ while $s_{12}(f)$ and $s_{21}(f)$ are complexly conjugated cross-spectral densities.

The other frequency-dependent quantities that characterize a bivariate random function of time (a bivariate time series) in the frequency domain are the coherence function

$$\gamma_{12}(f) = \frac{|s_{12}(f)|}{[s_{11}(f)s_{22}(f)]^{1/2}}, \tag{7.17}$$

the coherent spectrum

$$s_{11.2}(f) = \gamma_{12}^2(f)s_{11}(f), \tag{7.18}$$

and the complex-valued frequency response function

$$h_{12}(f) = s_{12}(f)/s_{22}(f) \tag{7.19}$$

with its gain factor

$$g_{12}(f) = |s_{12}(f)|/s_{22}(f) \tag{7.20}$$

and phase factor

$$\phi_{12}(f) = \tan^{-1}\{\mathrm{Im}[s_{12}(f)]/\mathrm{Re}[s_{12}(f)]\}. \tag{7.21}$$

The coherence function is dimensionless. Here, $\mathrm{Im}(A)$ and $\mathrm{Re}(A)$ are the imaginary and real parts of a complex-valued quantity A.

The spectral densities or spectra $s_{11}(f)$ and $s_{22}(f)$ describe the behavior of the output and input scalar processes in the frequency domain. Their estimates obtained for the same time series as a scalar quantity and as a scalar component of a multi-variate system may differ. The degree of discrepancy between them depends upon the system's complexity, time series length, and the orders of the scalar and multivariate autoregressive models.

The coherence function or coherence given with Eq. (7.17) satisfies the condition $0 \leq \gamma_{12}(f) \leq 1$. It is dimensionless, and it describes the degree of linear interdependence between the time series. It usually changes with frequency and can be regarded as a frequency-dependent cross-correlation coefficient between $x_{1,t}$ and $x_{2,t}$.

In the case of time-invariant random vectors, the amount of information contained in vector $x_{1,n}$ about another such vector $x_{2,n}$ (and vice versa) is

$$J = -\log(1 - r_{12}^2), \tag{7.22}$$

where r_{12} is the cross-correlation coefficient between the random variables $x_{1,n}$ and $x_{2,n}$.

When $\mathbf{x}_t = [x_{1,t}, x_{2,t}]'$ is a bivariate random function of time (a bivariate time series), the information rate $i(x_{1,t}, x_{2,t})$—the average amount of information per unit time contained in $x_{1,t}$ about $x_{2,t}$ and vice versa—is defined as

$$i(x_{1,t}, x_{2,t}) = -\int\limits_0^{f_N} \log[1 - \gamma_{12}^2(f)]df, \qquad (7.23)$$

where $f_N = 1/2\Delta t$ is the Nyquist frequency. These last two equations given in Gelfand and Yaglom (1957) constitute a mathematically strict explanation of why using the cross-correlation coefficient to describe relationships between time series is incorrect. As mentioned in the previous subsection, there is just one case when the correlation coefficient can serve as an indicator of interdependence between time series: it is when the time series $\mathbf{x}_t$ presents a white noise. However, in practical research, such models are to be the results of analyzing the time series by fitting to it AR models of different orders and making sure that the white noise is indeed the optimal model recommended by the majority of order selection criteria. Such cases are relatively rare in practice because of the sampling variability of estimates. To establish that the time series is a bivariate white noise one has to analyze it in the frequency domain to make sure that the spectra $s_{11}(f)$ and $s_{22}(f)$ do not change with frequency.

The information rate $i(x_{1,t}, x_{2,t})$ is defined as an average over the entire frequency range amount of information per unit time. Therefore, it can be used to characterize the degree of linear dependence between the time series: the interdependence becomes stronger if the information rate increases.

The two results given just above were obtained for Gaussian random vectors and Gaussian random processes but a Gaussian distribution is quite common in the Earth sciences and, besides, the authors of that classical work believe that their equations will hold in more general cases.

The coherent spectrum $s_{11.2}(f)$ shows the share of the output process's energy generated due to the linear dependence of the output upon the input. Its dimension coincides with the dimension of the spectrum $s_{11}(f)$. The coherent spectrum can also be used for evaluating the causality within multivariate time series.

The frequency response function $h_{12}(f) = s_{12}(f)/s_{22}(f)$ is complex-valued and its modulus—the gain factor $g_{12}(f)$—describes how the input process $x_{2,t}$ is transformed into the output $x_{1,t}$ within the frequency domain. Its dimension is the ratio of the output and input dimensions. It can be regarded as a frequency-dependent analog of regression coefficient.

The phase factor $\phi_{12}(f)$ shows the phase shift between the output and input, and it is usually expressed in radians. In a physically realizable system, the output lags the input. The time lag $\tau_{12}(f)$ at frequency f is defined as

$$\tau_{12}(f) = \phi_{12}(f)/2\pi f, \ f > 0. \tag{7.24}$$

Another frequency domain property of any multivariate stationary time series is the residual spectral density, which is obtained as the difference between the output and the coherent spectra:

$$n_{ii.j}(f) = s_{ii}(f) - s_{ii.j}(f), \ i, j = 1, 2. \tag{7.25}$$

The residual spectrum shows the behavior of the part of the output process which is not linearly related to the input time series. It does the same for the input process if $i = 2$ and $j = 1$. The quantity $n_{ii.j}(f)$ is also called the noise spectrum but one has to understand that it plays the role of noise only for the specific system that is being investigated and may generally play a dominant role. Examples are given in Sect. 8.2 and Chap. 13.

When the time series dimension M is higher than 2, the coherent spectral density is changed to the multiple coherent spectrum, which reflects the contribution of all input processes to the output $x_{1,t}$. The case of a trivariate time series is discussed in Chap. 14.

Computing the spectral characteristics of multivariate AR models through Eq. (7.15) and order selection criteria can be regarded as a multivariate version of the scalar maximum entropy method introduced by Burg (1967).

The autoregressive analysis of bivariate (and multivariate) time series is convenient to conduct in both time and frequency domains in an iterative manner for values of AR order from $p = 1$ through $p = p_{\max}$. The time domain version is given in the VAR package in Shumway and Stoffer (2017) or in MATLAB. However, those sources do not seem to allow one to calculate the spectral matrix through the time domain model. To the best of the author's knowledge, such packages are not available in either R or in the commercial software.

In concluding this section, consider an example of applying spectral analysis to a sample of a bivariate random process whose scalar components present a strictly linear transformation of each other.

Let the components $x_{1,t}$ and $x_{2,t}$ of the time series $\mathbf{x}_t = [x_{1,t}, x_{2,t}]'$ be

$$\begin{aligned} x_{1,t} &= \phi_1 x_{1,t-1} + x_{2,t} \\ x_{2,t} &= a_t \end{aligned} \tag{7.26}$$

where a_t is white noise and $|\phi_1| < 1$. These equations can be written as

$$x_{1,t} = (1 - \phi_1 B)^{-1} x_{2,t} \tag{7.27}$$

or

$$x_{2,t} = (1 - \phi_1 B) x_{1,t}. \tag{7.28}$$

Actually, we are treating the innovation sequence of an AR(1) time series as a component of a bivariate process. Equations (7.27) or (7.28) mean that the time series are related to each other through a linear operator. Consequently, the coherence function $\gamma_{12}(f)$ should be equal to unity. As follows from Eq. (7.26), the cross-correlation coefficient r_{12} between $x_{1,t}$ and $x_{2,t}$ is $\sqrt{1 - \phi^2}$. If $\phi_1 = 0.8$, the coefficient $r_{12} = 0.6$. These results obtained directly from Eq. (7.26) have been verified by analyzing a simulated bivariate time series (7.26) of length $N = 10^5$. The estimate of coefficient r_{12} was 0.6, and it was the maximum value of the entire cross-correlation function. All values of coherence function were equal to unity as it should be for any linearly dependent time series. In this specific case and generally, the cross-correlation function does not show that time series are strictly linearly related to each other, but the true relationship can be revealed through the coherence function.

7.1.3 Reliability of Autoregressive Estimates of Frequency-Dependent Quantities

All estimates of AR model parameters and spectral characteristics must be supplied with respective confidence intervals, which are computed through the variance of estimate errors. For the AR coefficients, such information is available in Box et al. (2015), in the multivariate time series software in R (Shumway and Stoffer, 2017) and in MATLAB. Variances of frequency-dependent quantities estimated with different nonparametric methods are given, for example, in Bendat and Piersol (1966, 2010) and, for scalar time series, in Percival and Walden (1993). However, the author is not aware of any publication that would provide respective information for estimates of frequency-dependent quantities obtained through AR models of multivariate time series. In this book, the confidence intervals for those estimates are approximate in the sense that the error variances of estimates are calculated under the assumption that they follow the equations given in Chaps. 8 and 9 of the Bendat and Piersol book (2010). The number of averages n_d for the estimates is defined as $N/M^2 p$, where M is the time series dimension ($M = 2$ in the bivariate case). This approach is M times more conservative than in the earlier work on the subject (Privalsky et al. 1987). According to the results recently obtained by the author, the frequency-dependent characteristics estimated through AR models of bivariate geophysical time series do not generally contain sharp peaks or troughs and are rather smooth. The Monte Carlo experiments with about a dozen such models showed that estimates of all spectral characteristics listed above have a low bias and the behavior of their variance does not seem to contradict the above-given assumption. Of course, those results are approximate. The statistical reliability problem for autoregressive estimates of frequency-dependent functions in the multivariate case does not seem to have been resolved but at this time there is just no other way but to use those approximate results. Similar to what has been done in previous chapters, the confidence intervals for estimates of all spectral characteristic in the multivariate case will always be given for the 90% confidence bounds (or at a significance level 0.1).

7.2   Granger Causality and Feedback

The term "causality" had been introduced into time series analysis by the Nobel Prize laureate in economics Sir Clive Granger (Granger 1963). Generally, causality between time series $x_{1,t}$ and $x_{2,t}$ means that the time series $x_{2,t}$ affects in some way the time series $x_{1,t}$ and/or vice versa. A different definition of causal time series given in Shumway and Stoffer (2017) is more general, and it coincides with the concept of physical realizability and stationarity. The Granger causality has a rather specific meaning: the time series $x_{2,t}$ "causes" the time series $x_{1,t}$ if the error variance of extrapolation of $x_{1,t}$ at the unit lead time as a member of a bivariate time series $\mathbf{x}_t = [x_{1,t}, x_{2,t}]'$ is smaller than the error variance of extrapolation of $x_{1,t}$ at the unit lead time as a scalar time series, without taking $x_{2,t}$ into account. If the condition of causality is fulfilled for both processes $x_{1,t}$ and $x_{2,t,}$ then, according to C. Granger, the bivariate time series presents a feedback system.

The distinguished author of this causality idea considered it important, at least, for applications in econometrics, and discussed it in a series of publications, including the classical book on spectral analysis (Granger and Hatanaka 1964) and his Nobel Prize lecture (Granger 2004). Other papers on the subject include Granger (1969, 1980), and Granger and Lin (1995).

The Granger causality criteria for the time series $\mathbf{x}_t = [x_{1,t}, x_{2,t}]'$ are

$$c_{12} = (1 - v_{12}/v_{11}) \tag{7.29}$$

and

$$c_{21} = (1 - v_{21}/v_{22}), \tag{7.30}$$

where v_{12} and v_{11} are the variances of extrapolation errors of $x_{1,t}$ at the unit lead time with and without the information about $x_{1,t}$ and similarly for $x_{2,t}$. Thus, $v_{11} = \sigma_a^2$ and $v_{12} = P_{11}$ are the variances of innovation sequences in the scalar AR model (3.3) and in the bivariate **AR** model (7.4), respectively. Therefore, the information needed for calculating the Granger causality and feedback criteria is available in the bivariate and scalar autoregressive models of the time series.

The feedback strength, according to C. Granger, is

$$s_{12} = c_{12}c_{21} \tag{7.31}$$

and, obviously, $s_{12} = s_{21}$. All three criteria lie between 0 and 1.

In his book with Hatanaka (1964), C. Granger shows that his causality and feedback criteria are related to the Gelfand and Yaglom information rate $i(x_{1,t}, x_{2,t})$

as

$$\log(1 - c_{12}) + \log(1 - c_{21}) = -i(x_{1,t}, x_{2,t}) \tag{7.32}$$

and

$$1 + s_{12} - c_{12} - c_{21} = \exp[-i(x_{1,t}, x_{2,t})]. \tag{7.33}$$

The relation of Granger causality to the Gelfand and Yaglom information rate and, therefore, to the coherence function [see Eq. (7.23)] means that the concept of causality belongs to theory of random processes and information theory and, consequently, the time series are supposed to be predicted in accordance with the Kolmogorov–Wiener theory of extrapolation. As the Gelfand–Yaglom information rate is defined through the coherence function, it also means that Granger causality must be investigated within both time and frequency domains.

It should be stressed that the Granger causality criteria are based upon time series extrapolation; it means that if the quality of extrapolation is low, there will be no Granger causality and no Granger feedback even if the time series are closely interdependent. The simplest example of such interdependence is a bivariate white noise $\mathbf{x}_t = [a_{1,t}, a_{2,t}]'$. If the cross-correlation coefficient ρ_{12} between $a_{1,t}$ and $a_{2,t}$ is not zero, the components affect each other; in particular, the contribution of the second component to the variance of the first component is $\rho_{12}^2 P_{11}$ and it can be quite substantial if the cross-correlation coefficient ρ_{12} is high. Yet, both components remain unpredictable so that Granger causality and feedback do not exist in this case. The Granger causality definition does not take the correlation coefficient ρ_{12} into account explicitly, which may lead to a discrepancy between Granger's definitions of causality or feedback and the direct estimation of the information rate through the coherence function estimate given with Eq. (7.23).

At climatic timescales, the degree of time series predictability is often defined by the degree of prevalence of low frequencies in the time series spectrum. External nature-caused factors including teleconnection strength do not generally play a significant role because they usually occur at frequencies where the spectral density of the process is small. Because of this, the issue of Granger causality may not be so important for processes that characterize climate, such as spatially averaged time series of surface temperature (global, hemispheric, oceanic, and terrestrial). It means that applying Granger's criteria to real time series may be disappointing in some cases. Firstly, the ratios v_{12}/v_{11} and v_{21}/v_{22} are often close to unity in climatology and related sciences. Therefore, the values of causality criteria turn out to be quite small and the strength of Granger feedback close to zero. Secondly, due to the sampling variability of estimates and the role of order selection criteria, the orders p of the bivariate and scalar models $\mathbf{AR}(p)$ and $AR(p)$ may be different and the estimate of extrapolation error variance v_{12} may be larger than the error variance v_{11}; that is, the scalar extrapolation will have the prediction error variance smaller than for the bivariate extrapolation. Then, the criterion c_{12} will be negative and one will have to admit that the results of bivariate analysis are worse, in the predictability sense, than in the scalar case. In other words, there may be cases when the Granger causality and feedback criteria do not provide useful information about causality and feedback in Granger's sense. Some help in determining causality and feedback can be provided by using the time domain Eqs. (7.12) and (7.13).

In concluding this section, one should mention the approach to determining dependence and causality between time series suggested in a series of publications in a physical journal (Liang 2014, 2015, 2016). The author of the approach explains relations between time series through cross-correlation coefficients. According to X. Liang, a significant cross-correlation between time series means that they are related to each other but it does not mean that there is any causality between them, that is, that one time series "causes" the other one. This result is incorrect because the dependence between time series and causality may exist irrespective of the cross-correlation coefficient. It has been shown in this chapter that one time series can be completely and linearly dependent upon another time series (i.e., caused by it) even if the cross-correlation between them is zero. Such situation is rather trivial for information theory.

7.3 On Properties of Software for Analysis of Multivariate Time Series

The software required for analysis of multivariate time series is relatively cumbersome. It is considerably more complicated than what is needed for calculating cross-correlation coefficients and regression equations, which are still used traditionally in Earth and solar sciences for finding teleconnections and for reconstructing time series. In addition to the basic statistical information such as mean values, variances, matrix covariance and correlation functions, the output results should include information about time domain models (which are always autoregressive in this book) and about time series behavior in the frequency domain. This should be done for each $\mathbf{AR}(p)$ model for orders from $p = 1$ through the maximum order $p = p_{\max}$ that should be selected in accordance with the time series length N and the dimension M. Thus, in the bivariate case and $N = 200$, the recommended maximum order $p_{\max}$ should be selected in such a way that the number of autoregressive coefficients that need to be estimated would be an order of magnitude smaller that the time series length. It means that if $M = 2$ (bivariate time series) and $N = 200$ the recommended maximum order should not exceed 5. This computationally convenient approach seems to be used in Shumway and Stoffer (2017) in their package VAR.

In accordance with what has been given in this chapter, the time domain information about an $\mathbf{AR}(p)$ model should include estimates of matrices of autoregressive coefficients (7.6) and the covariance matrix of innovation sequence (7.7). All these estimates should be accompanied with respective error variances.

The estimates of the covariance matrix (7.9) are needed both for the general idea about the time series behavior in the time domain and specifically for using Eqs. (7.12) and (7.13) to describe the time domain dependence between the time series components.

The information about time series properties in the frequency domain should be given for each order chosen by order selection criteria. The frequency domain statistics consist of the following functions:

- spectral densities,
- coherent spectra,
- coherence functions,
- gain factors,
- phase factors, and, if required,
- time lags, and
- residual (noise) spectrum.

In the multivariate case, when $M > 2$, the amount of required information significantly increases because it has to include multiple and partial coherence functions, multiple and partial coherent spectra, the noise spectra, and frequency response functions (its gain and phase factors) for each tract of the system. It may also include time lags (7.24) for each input process. Examples of trivariate time series analysis in time and frequency domains are given in Chap. 14.

The frequency domain analysis is convenient to conduct through the results of time domain analysis within the same software package. The estimates of coherence functions obtained simultaneously with the time domain information would allow one to immediately decide whether the scalar components of a bivariate (or multivariate) time series are interdependent. An early version of this author's software that had been published for free 25 years ago (Privalsky and Jensen 1995b) satisfies most of the above-listed requirements.

A recent book by W. Wei (2019) contains a package in R for frequency domain analysis of multivariate time series. It can be used for spectral analysis of bivariate or multivariate time series but the results do not contain confidence limits for estimates of all frequency-dependent functions. Further comments on Wei's book are given in Chap. 14.

References

Bendat J, Piersol A (1966) Measurement and analysis of random data. Wiley, New York
Bendat J, Piersol A (2010) Random data. Analysis and measurements procedures, 4th edn. Wiley, Hoboken
Bhowmik R, Sankarasubramanian B (2019) Limitations of univariate linear bias correction in yielding cross-correlation between monthly precipitation and temperature. Int J Climatol 39:4479–4496
Box G, Jenkins M, Reinsel G, Liung G (2015) Time series analysis. Forecasting and control, 5th edn. Wiley, Hoboken

Burg JP (1967) Maximum entropy spectral analysis. Paper presented at the 37th Meeting of Society for Exploration Geophysics. Oklahoma City, OK, October 31, 1967

Christiansen B, Ljungkvist F (2017) Challenges and perspectives for large-scale temperature reconstructions of the past two millennia. Rev Geophys 55(1):40–97

Gelfand I, Yaglom A (1957). Calculation of the amount of information about a random function contained in another such function. Uspekhi Matematicheskikh Nauk 12:3–52. English translation: Am Math Soc Trans Ser 2(12):199–246, 1959

Granger C (1963) Economic processes involving feedback. Inf Control 6:28–48

Granger C, Hatanaka M (1964) Spectral analysis of economic time series. Princeton University Press, Princeton

Granger C (1969) Investigating causal relations by econometric models and cross-spectral methods. Econometrica 37:424–438

Granger C (1980) Testing for causality: a personal viewpoint. J Econ Dyn Control 2:329–352

Granger C (2004) Time series analysis, cointegration, and applications. Am Econ Rev 94:421–425

Granger C, Lin J-L (1995) Causality in the long run. Economet Theor 11:530–537

Jin C, Liu J, Wang B et al (2019) Decadal variations of the East Asian summer monsoon forced by the 11-year insolation cycle. J. Climate 32:2735–2745

Khan A, Chen F, Ahmed M, Zafar M (2020) Rainfall reconstruction for the Karakoram region in Pakistan since 1540 CE reveals out-of-phase relationship in rainfall between the southern and northern slopes of the Hindukush-Karakorum-Western Himalaya region. Int J Climatol 40:52–62

Kumar V, Melet A et al (2018) Reconstruction of local sea-levels at south west Pacific islands—a multiple linear regression approach (1988–2014). J Geophys Res Oceans 123(2):1001–1015

Liang X (2014) Unraveling the cause-effect relation between time series. Phys Rev E 90:052150-1–052150-11

Liang X (2015) Normalizing the causality between time series. Phys Rev E 92:022126-1–022126-7

Liang X (2016) Information flow and causality as rigorous notions ab initio. Phys Rev E 94:052201-1–052201-28

Percival D, Walden A (1993) Spectral analysis for physical applications. Cambridge University Press, Cambridge

Privalsky V, Protsenko I, Fogel G (1987) The sampling variability of autoregressive spectral estimates for two-variate hydrometeorological processes. In: Proceedings of 1st World Congress of the Bernoulli Society on Mathematical Stat. Theory and Applications, vol 2, VNU Science Press, Utrecht, pp 651–654

Privalsky V, Jensen D (1995a) Assessment of the influence of ENSO on annual global air temperature. Dynam Atmos Oceans 22:161–178

Privalsky V, Jensen D (1995b) Time series analysis package. Utah Climate Center, Utah State University, Logan, UT

Privalsky V, Yushkov V (2018) Getting it right matters: climate spectra and their estimation. Pure Appl Geoph 175:3085–3096

Reinsel G (2003) Elements of multivariate time series analysis, 3rd edn. Springer, New York

Salinger M, Fitzharris B, Chinn T (2019) Atmospheric circulation and ice volume changes for the small and medium glaciers of New Zealand's Southern Alps mountain range 1977–2018. Int J Climatol 39:4274–4287

Schweiger A, Wood K, Zhang J (2019) Arctic sea ice volume variability over 1901–2010: A model-based reconstruction. J Clim 32:4731–4752

Shumway R, Stoffer D (2017) Time series analysis and its applications. Springer, Heidelberg

Simpson I, Yeager S et al (2019) Decadal predictability of late winter precipitation in western Europe through an ocean–jet stream connection. Nat Geosci 12:613–619

Tan L et al (2019) Rainfall variations in central Indo-Pacific over the past 2,700 y. PNAS 27(116):17201–17206

Wei W (2019) Multivariate time series analysis and applications. Wiley, Hoboken

Yaglom AM (1987) Correlation theory of stationary and related functions. Basic resutls. Springer, New York

Yang R, Gui S, Cao J (2019) Bay of Bengal-East Asia-Pacific Teleconnection in Boreal Summer. J Geophys Res-Atmos 124:4395–4412

Zhang T, Lou B, et al (2019) A 256-year-long precipitation reconstruction for northern Kyrgyzstan based on tree-ring width. Int J Climatol https://doi.org/10.1111/1751-2980.12715

Chapter 8
Teleconnection Research and Bivariate Extrapolation

Abstract This chapter is devoted to studying teleconnections. Teleconnections at a sampling rate of one year are between the components of the ENSO system (SOI and NINO) and between NINO and nine time series of spatially averaged annual surface temperature (AST): the globe, hemispheres, global and hemispheric oceans, and land. A meteorological teleconnection is between the components of MJO; its bivariate predictability is briefly analyzed in accordance with KWT. The system SOI/NINO is a strong teleconnection controlled by the innovation sequences which are closely correlated with each other. The system's components are almost independent of their past and have spectra with smooth maxima at about 0.22 cpy; their predictability is low. ENSO is shown to affect AST being responsible for about 10% of the total variance of spatially averaged and detrended temperature and for 35–50% in the vicinity of ENSO's natural frequency. The coherence can be up to 0.9, the change in NINO by 1 °C causes a response of about 0.1 °C in AST; the delay between NINO and AST is from 0.4 to 0.7 year. The feedback and Granger causality are weak. If the trend is regarded as nature-caused, the NINO contribution to surface temperature is negligibly small.

This chapter describes the method of teleconnection research and contains ten examples of teleconnection analysis within the Earth's climate system at a climatic sampling rate of one year and an example of teleconnection and extrapolation analysis of a bivariate process at a meteorological timescale of one day. The ENSO is studied as a teleconnection system consisting of Southern Oscillation in the atmosphere and an oceanic component NINO represented with sea surface temperature in equatorial Pacific. Teleconnections within the frequency band close to ENSO's natural frequency are shown to exist between NINO and the annual surface temperature averaged over the entire globe, its hemispheres, oceans, and land. The MJO example includes analysis of the connection between its components and contains an extrapolation analysis of MJO as a bivariate time series in accordance with the Kolmogorov–Wiener theory.

© Springer Nature Switzerland AG 2021
V. Privalsky, *Time Series Analysis in Climatology and Related Sciences*,
Progress in Geophysics, https://doi.org/10.1007/978-3-030-58055-1_8

8.1   Example 1. The ENSO Teleconnection

Traditionally, ENSO behavior is studied within the framework of classical mathematical statistics and, consequently, without the knowledge of its time domain properties such as the system's memory and statistical predictability and without its spectral analysis as a bivariate time series including the functions described in Chap. 7. Earlier, a stochastic model of ENSO as a teleconnection system has been suggested and analyzed in time and frequency domains by Privalsky and Muzylev (2013); among other issues, this section presents an update of some results given in that publication.

The El Niño-Southern Oscillation (ENSO) system is a typical teleconnection: the sea surface temperature in the western equatorial Pacific closely related to the Southern Oscillation Index (SOI)—transformed differences of atmospheric pressure between Tahiti Island and Darwin (Australia). The interdependence between these two climate variables at three different locations on the globe matches the concept of teleconnection. In this section, we will discuss the teleconnection between the time series of annual SOI and annual sea surface temperature in the ENSO area NINO3.4 (5° S–5° N, 120° W–170° W). The data Web sites are #1 and #2 in Appendix to this chapter. The time series are shown in Fig. 8.1. In what follows, the SST data will be called NINO.

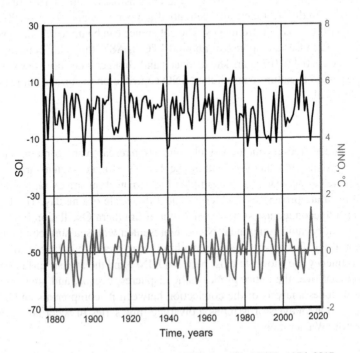

Fig. 8.1 Annual Southern Oscillation Index and SST anomalies NINO, 1876–2017

Judging by the behavior of the time series, there is no need in this case to test the stationarity assumption. Both time series pass the Gaussianity test so that the bivariate time series SOI/NINO can be regarded as ergodic. The cross-correlation coefficient between SOI and NINO time series is -0.81. At all other lags, the cross-correlation function is close to zero (Fig. 8.2), which probably means that the absolute value of the cross-correlation coefficient $r_{12}(0) \approx -0.81$ is high due to the high cross-correlation between the innovation sequences $a_{1,t}$ and $a_{2,t}$.

The **AR**(2) model chosen for the ENSO bivariate time series is

$$x_{1,t} \approx \mathbf{0.16}x_{1,t-1} - \mathbf{1.07}x_{2,t-1} + \mathbf{0.06}x_{1,t-2} + 4.37x_{2,t-2} + a_{1,t}$$
$$x_{2,t} \approx \mathbf{0.02}x_{1,t-1} + 0.43x_{2,t-1} + \mathbf{0.002}x_{1,t-2} - 0.30x_{2,t-2} + a_{2,t}, \qquad (8.1)$$

where $x_{1,t}$ and $x_{2,t}$ are SOI (the output) and NINO (the input), respectively. The coefficients shown in bold are statistically insignificant. The choice of SOI as the output is tentative; the final decision about which quantities should be selected as the input and output has to be made in the process of further analysis.

As seen from Eq. (8.1), the output process (SOI) is practically independent of its own past values (the AR coefficients at $x_{1,t-k}$ are statistically insignificant) and

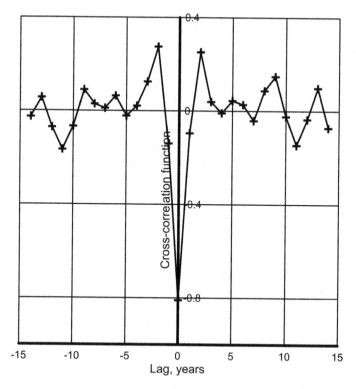

Fig. 8.2 Cross-correlation function between SOI and NINO

Table 8.1 Summands of SOI and NINO variances

Source	SOI, %	NINO, %
From SOI	1	−1
From NINO	11	16
From $a_{1,t}$	31	55
From $a_{2,t}$	57	30

depends upon the value of the input (NINO) that has occurred two years ago. The input process depends upon its own past but not upon the output. This is a system with an open feedback loop.

The covariance matrix of the innovation sequence $\mathbf{a}_t$ is

$$\mathbf{P_a} \approx \begin{bmatrix} 41.8 & -2.74 \\ -2.74 & 0.28 \end{bmatrix}. \tag{8.2}$$

According to this, the cross-correlation coefficient between the components of the innovation sequence is high: $\rho_{12} \approx -0.81$. It coincides with the cross-correlation coefficient between SOI and NINO. Thus, the strong interdependence between SOI and NINO occurs due to the high correlation between the components of the bivariate white noise $\mathbf{a}_t$.

Results of calculations with Eqs. (7.12) and (7.13) given in Table 8.1 show that the variance σ_1^2 of SOI is practically independent of SOI's past and that 11% of the variance is generated by the past values of NINO. The innovation sequence creates 88% of SOI's variance of which 57% is provided by the part of the innovation sequence $a_{1,t}$ linearly related to $a_{2,t}$; the remaining 31% is contributed by the part of $a_{1,t}$ not related to $a_{2,t}$. With 88% of its variance generated by the innovation sequences, the SOI time series is close to a white noise.

The input time series NINO with 85% of its variance consisting of contributions from the innovation sequences is also close to white noise; more than a half (55%) of the variance P_{22} is generated by $a_{1,t}$. Past values of NINO provide about 16% of its variance while the SOI contribution is negligibly small. Overall, the time series SOI/NINO is close to a bivariate white noise because its dependence upon the past is weak and the innovation sequence plays a dominant role in its behavior. These features make the bivariate ENSO time series practically unpredictable at climatic timescales.

This close resemblance to white noise is responsible for the lack of persistence so that the probability of El Niño or La Niña phenomena (an increase or drop of annual temperature by not less than 0.5 °C above or below the mean, respectfully) occurring several years in a row is small. An experiment with a simulated SOI/NINO time series of length $N = 10^5$ showed that the probabilities of occurrence are about 0.1, 0.04, and 0.006 for one, two, and three years in a row both for El Niño and La Niña. At $\Delta t = 1$ year, the longest sequence of years with El Niño's or La Niña's according to this experiment, could be 6 years with probability 4×10^{-5} (once in twenty-five millennia). The statistical predictability of time series of annual SOI and NINO is about equally low for the bivariate **AR**(2) and scalar AR models of the processes.

The improvement of predictability for the bivariate model as compared to the scalar cases is just a few percent so that the Granger causality and feedback are practically nonexistent for this bivariate time series. At the same time, the influence of time series components upon each other is quite strong in the **AR**(2) model due to the high cross-correlation within its innovation sequence. According to Granger's formulae given in Sect. 7.6, the Gelfand–Yaglom information rate should be 0.1 while the direct calculation through the coherence function estimate is 0.6. This discrepancy happens because the high information rate from SOI to NINO and vice versa occurs due to a high coherence between them, but the high coherence results from the strong correlation between the unpredictable innovation sequences. This is an example of the lack of Granger causality in spite of a close interdependence of time series components.

The spectra of SOI and NINO (Fig. 8.3) are not monotonic and contain smooth maxima at $f \approx 0.21$ cpy, which is very close to the natural frequency of ENSO system expressed with Eq. (8.1). The minor role of the spectral maxima is seen from the fact that in both cases the confidence intervals for the spectral estimates are wide enough to draw a monotonic line between the confidence limits. This is equally true for the AR (MEM) and MTM spectral estimates.

The coherence function (Fig. 8.4a) is statistically significant at all frequencies and its maximum value is 0.92. Another result of the high coherence is that the coherent spectral density (Fig. 8.4b) explains 50–80% of the SOI spectrum. Such situation seems to be unusual at climatic timescales because in this case the coherence attains its highest values at the frequency band where the spectral density also reaches its maximum. The total contribution of the linear dependence between the components to the SOI variance expressed as the ratio of the integral of the SOI's coherent spectrum

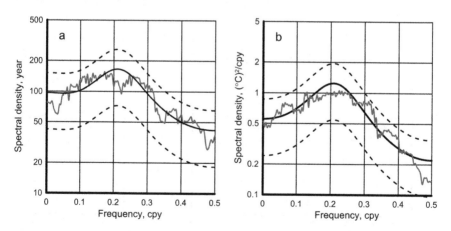

Fig. 8.3 AR (black) and MTM (gray) estimates of SOI (**a**) and NINO (**b**) spectra

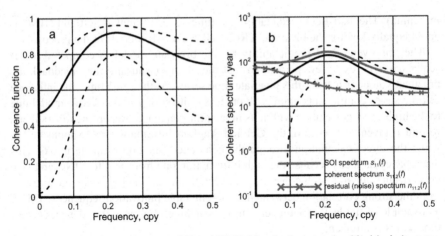

Fig. 8.4 Coherence function between SOI and NINO (**a**) and coherent and residual (noise) spectra (**b**)

to the integral of SOI's spectral density is over 60% and the residual spectrum also shown in Fig. 8.4b is responsible for more than 30% of the SOI spectrum.

The linear dependence between SOI and NINO caused by the innovation sequences plays a dominant role in this bivariate stochastic system. This also means that statistical forecasting of annual values of SOI and/or NINO as scalar processes or as components of a bivariate process will always have large error variance even at the unit lead time. In other words, the bivariate ENSO system has low statistical predictability at climatic timescales.

The transformation of NINO into SOI occurs with an amplification coefficient from 6 to 11 index units per 1 °C of NINO change; the highest values occur at frequencies over 0.2 cpy (Fig. 8.5a). In the system where NINO and SOI present

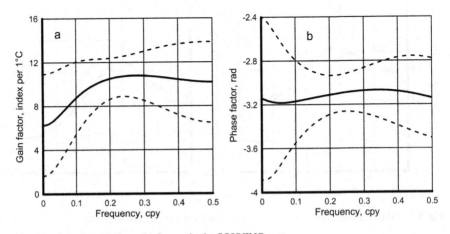

Fig. 8.5 Gain (**a**) and phase (**b**) factors in the SOI/NINO system

the output and input, the change of SOI by one index unit will cause the sea surface temperature to change by 0.06–0.08 °C.

The phase factor (Fig. 8.5b) shows that the shift between the components is always close to $-\pi$ which explains the negative correlation between the time series. In its turn, the high negative correlation between time series occurs due to the high negative correlations between the innovation sequence components and due to the large variances of those components. The deviations from $-\pi$ are small and the confidence interval for the phase factor estimate includes a constant phase shift equal to $-\pi$ between SOI and NINO over the entire frequency band.

The entire ENSO system is controlled by the bivariate noise $\mathbf{a}_t$ with a phase shift between its components equal to $-\pi$. If SOI were defined as the differences of atmospheric pressure between Darwin and Tahiti, the phase shift would have been zero. The phase shift and, consequently, the time lag within a bivariate white noise system can be very close to zero and this is what is observed here if the sign of SOI is changed. This means that the choice of the input and output in the SOI/NINO system is not important because of the dominant influence of the white noise.

Conclusions about the behavior of the SOI/NINO stochastic system in the time domain can be formulated in the following manner:

- interannual variations of the Southern Oscillation Index are practically independent of its past values and almost 90% of SOI's variance is generated by the white noise innovation sequence.
- the rest of the SOI variance is generated by the past values of NINO.
- about 15% of the NINO variance is formed through the dependence upon its own past values while the innovation sequence $\mathbf{a}_t$ is responsible for approximately 85%.
- the high correlation between the innovation sequences results in substantial (55–57% of respective variances) indirect influence of the time series components upon each other.
- to a large extent, the system behaves as a bivariate white noise with closely cross-correlated components.
- granger causality and feedback strength are very low though the time series are strongly interdependent.

The analysis also reveals that statistical predictability of SOI and NINO is better in the bivariate case but the differences from the scalar case are small.

The frequency domain analysis shows that

- the spectral maxima at 0.21–0.22 cpy are very smooth, which means that the year-to-year variations of ENSO do not contain any visible cyclicity.
- the coherence between SOI and NINO is high, reaching 0.92 in the vicinity of the spectral maxima at about 0.22 cpy and staying above 0.70 at all frequencies higher than 0.1 cpy.
- the coherent spectrum that shows the role of the linear dependence between the components is responsible for over 60% of the SOI and NINO spectral densities varying over the entire frequency band from 50% to 80%.

- a change of NINO by 1 °C causes a change of SOI by 6-11 index units.
- the absolute value of the phase factor is close to π at all frequencies confirming that the system is controlled by the bivariate white noise so that SOI and NINO can be equally regarded as either the input or the output of the system.

This example illustrates the important role of correlation between the innovation sequences: if it is high, the predictability, Granger causality and feedback will be low in spite of the strong interdependence between the time series. The major points in this example are that at climatic timescales (starting from year-to-year variability) the ENSO system is close to a white noise, contains no regular oscillations and has low statistical predictability in both scalar and bivariate cases.

8.2 Example 2. Teleconnections Between ENSO and AST

The ENSO is believed to influence many other processes on the globe but those teleconnections are usually studied through the cross-correlation coefficient (e.g., He et al. 2018; Yeh et al. 2018) or for individual seasons (e.g. Yuang and Wang 2019). The reliability of such analyses is doubtful. In the example below, the role of ENSO in the behavior of nine major climatic time series is analyzed at interannual timescales within the framework of theory of random processes and information theory. A general goal here is to test the hypothesis of ENSO's influence upon annual surface temperature averaged over large parts of the globe given in Privalsky and Yushkov (2018). A more specific goal is to get detailed quantitative information in both time and frequency domains about the influence of ENSO upon surface temperature obtained by averaging over the oceans and land.

8.2.1 Time Domain Analysis—ENSO and Spatially Averaged Temperature

The effect of ENSO elements upon global temperature has been studied within the framework of theory or random processes by Privalsky and Jensen (1995) using SOI as the input process. In particular, it has been shown that though the total contribution of SOI to the annual global temperature was about 10%, the contribution within the frequency band from 0.08 cpy through 0.25 cpy was at least 30%. Some of the conclusions made in that publication need to be updated and amended by analyzing the previously unavailable data sets—the nine time series of spatially averaged annual surface temperature (AST). Respective data were taken from the web site of the University of East Anglia (reference #3 in Appendix) while ENSO is represented not with the SOI time series as in the above-mentioned publication but with the same NINO time series as in the previous example. The annual NINO time series and

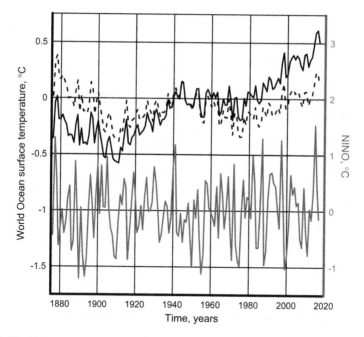

Fig. 8.6 World Ocean surface temperature (black, with and without the trend) and sea surface temperature in the ENSO area NINO (gray), 1876–2017

the time series of surface temperature averaged over the World Ocean are shown in Fig. 8.6 as an example of the nine bivariate time series studied in this section.

In what follows, the surface temperature time series are regarded as outputs $x_{1,t}$ of respective linear stochastic systems with the NINO time series $x_{2,t}$ as the input. The abbreviations for the AST time series including global, hemispheric, oceanic, and terrestrial temperatures are GLOBE, NH, SH, OCEAN, ONH, OSH, and LAND, LNH, LSH.

Similar to many other climate indices, the NINO time series does not show any significant trend, but all AST time series contain strong linear trends. Therefore, the first question to answer is whether the trend should be deleted. For this purpose, the spectral estimates of the time series prior to and after trend deletion should be compared with each other. The way to do it is described in Chap. 4. As seen from Fig. 8.7, the low-frequency share of global temperature spectrum is higher at frequencies below 0.020–0.025 cpy, that is, at timescales longer than 40–50 years.

Such scales are comparable to the time series length of 142 years and cannot be analyzed reliably. Suppressing them will not distort most results at higher frequencies. Besides, judging by the shape of the spectra in Fig. 8.7, the response of the spatially averaged temperature to ENSO can be expected at frequencies higher than 0.1 cpy so that trend deletion would hardly affect the behavior of surface temperature at those frequencies in a significant manner. If the trend is not deleted, it will dominate all other properties of surface temperature data. In particular, the variance

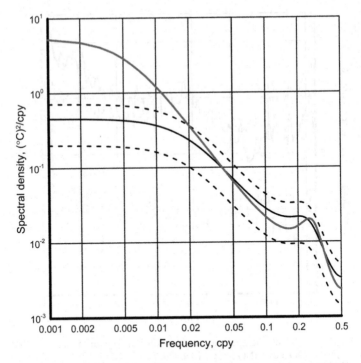

Fig. 8.7 Spectra of OCEAN time series with and without linear trend (gray and black)

of all time series will be heavily (up to 90%) dependent upon their own past values so that contributions of ENSO will be negligibly small. According to Fig. 8.7, the role of NINO at frequencies higher than 0.05 cpy will not be distorted by the presence of the trend. Yet, putting aside the reasons for the existence of this strong trend, we will regard the time series without the trend as generated by purely natural variations of surface temperature. Therefore, the trend has been deleted from all time series of AST.

As seen from Fig. 8.8, the correlation between the globally averaged oceanic temperature (OCEAN) and NINO is low; its maximum value is the cross-correlation coefficient $r_{12}(0)$, which amounts to 0.36. Out of the nine cases, the absolute maximum of all cross-correlation functions was 0.42 and it has been reached for the system LSH/NINO which differs rather significantly from all other systems. The cross-correlation coefficient $r_{12}(0)$ presents the maximum value of the cross-correlation function for the entire globe, hemispheres, and oceans and moves to $r_{21}(1)$ for the systems containing terrestrial components LAND, LNH, and LSH. Seemingly, the influence of ENSO upon the terrestrial AST takes more time.

It should also be mentioned that the level of statistical significance for the cross-correlation coefficient estimate traditionally used in climatology and related sciences is incorrect because it is always calculated for the number of variables in the data

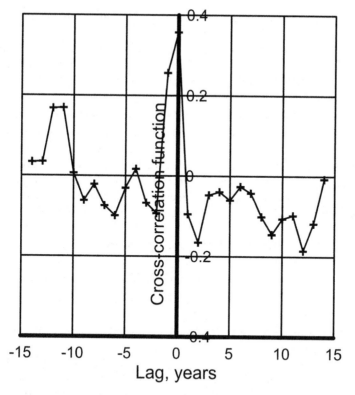

Fig. 8.8 Cross-correlation function between OCEAN and NINO

set as if the set consisted of mutually uncorrelated random variables. This is by no means acceptable for time series in general and specifically for all AST time series.

As shown in Chap. 7, the cross-correlation coefficient or any other individual value of cross-correlation function has no special meaning for multivariate time series; on top of that, its values in this case are always low. If one were to follow the traditional approach to teleconnection research, the conclusion about the effect of ENSO upon the global temperature would have been discouraging. The bivariate time series are studied here in accordance with the basics of multivariate time series analysis given in Chap. 7 so that information about possible teleconnection systems becomes available simultaneously for both time and frequency domains.

For NINO and the time series of annual surface temperature, all nine estimates of coherence function are statistically significant at frequencies above 0.1 cpy and may achieve 0.7–0.9 in a wide frequency range: at least from 0.20 cpy through 0.4 cpy. They show that the annual temperature over the nine major regions of the Earth surface, from global to southern hemisphere land is closely related to NINO—the oceanic component of ENSO. A more detailed analysis of coherence functions will be given in Sect. 8.2.2.

The analyses of these teleconnections should begin with the time domain properties of respective bivariate systems.

The optimal AR models for all nine bivariate time series turned out to be an **AR**(2). It has been chosen by all order selection criteria in practically all cases. The model for the GLOBE/NINO time series is

$$
\begin{aligned}
x_{1,t} &\approx 0.53x_{1,t-1} + 0.07x_{2,t-1} + 0.31x_{1,t-2} - 0.10x_{2,t-2} + a_{1,t} \\
x_{2,t} &\approx -1.10x_{1,t-1} + 0.40x_{2,t-1} + 0.90x_{1,t-2} - 0.32x_{2,t-2} + a_{2,t}.
\end{aligned} \tag{8.3}
$$

Here $x_{1,t}$ and $x_{2,t}$ are the GLOBE and NINO temperatures, respectively, and $\mathbf{a}_t = [a_{1,t}, a_{2,t}]'$ is the innovation sequence. All eight coefficients in Eq. (8.3) are statistically significant at the usual confidence level 0.9. This statement is true for almost all other estimates of AR coefficients in the bivariate models discussed in this section.

The covariance matrix of the white noise sequence $\mathbf{a}_t$ in Eq. (8.3) is

$$
\mathbf{P_a} \approx \begin{bmatrix} 0.0083 & 0.022 \\ 0.022 & 0.270 \end{bmatrix}, \tag{8.4}
$$

which corresponds to a cross-correlation coefficient $\rho_{12} \approx 0.46$ between the innovation components.

The bivariate **AR**(2) models for all other time series differ from the GLOBE/NINO model by moderate changes in estimates of **AR** coefficients, covariance matrices $\mathbf{P_a}$, coherence functions, etc., but qualitatively the results are quite similar for all models.

With the GLOBE and NINO variances, σ_1^2 and σ_2^2 equal to 0.025 (°C)^2 and 0.320 (°C)^2, the ratios $P_{11}/\sigma_1^2 \approx 0.33$ and $P_{22}/\sigma_2^2 \approx 0.84$, which means that the innovation sequence $a_{1,t}$ is relatively unimportant in the behavior of annual global temperature while variations of NINO are dominated by the white noise $a_{2,t}$. Variations of annual surface temperature averaged over the entire planet present a highly persistent random process while the oceanic component of ENSO is rather close to white noise. These results agree with the information about properties of spatially averaged temperature and NINO given in Privalsky and Yushkov (2018). The new result here is that the **AR**(2) model of the GLOBE/NINO bivariate time series describes a system with a closed feedback loop: the two processes affect each other. The feedback, including its strength, will be studied in what follows.

The variances of annual surface temperature averaged over large areas of Earth's surface vary in the range from 0.018 (°C)^2 to 0.077 (°C)^2 and, quite understandably, the variance is smaller for the oceanic areas (Table 8.2). The temperature variations heavily depend upon their values during the previous two years in all but one cases. The contributions from ENSO—the subject of this study—are small and stay between 3% and 11% while the effect of respective innovation sequences amounts to 35–45% and it is caused by the NINO's innovation sequence $a_{2,t}$.

The contributions of NINO—the oceanic component of ENSO—to variations of spatially averaged temperature show its role and, through it, the role of ENSO as

Table 8.2 Variance summands for spatially averaged annual temperature

Region	Variance, $(°C)^2$	From itself, %	From NINO, %	From $a_{1,t}$, %	From $a_{2,t}$, %
Globe	0.025	57	9	8	26
NH	0.041	59	6	4	31
SH	0.019	44	11	13	32
Ocean	0.020	60	7	12	21
ONH	0.030	69	3	8	20
OSH	0.018	55	5	8	32
Land	0.051	48	10	1	40
LNH	0.077	49	7	1	44
LSH	0.030	20	22	9	49

the input component of respective teleconnection systems. With one exception, the NINO's contribution does not exceed 11%, which explains the low cross-correlation coefficients between NINO and the other nine time series.

The exception is the southern hemisphere's terrestrial temperature (LSH). Its variance is much smaller than over the entire land surface (LAND) and over the northern hemisphere's land (LNH). Its past does not play a major role while the dependence upon NINO's past is unusually strong. Moreover, the role of the white noise in the LSH model is the highest among all nine bivariate systems and it attains almost 60%. The reasons for these unusual properties are not clear. They may be related to the relatively small area taken by land in the southern hemisphere or to a less reliable data set.

The ENSO contribution to the variance of AST within the frequency band close to the ENSO natural frequency of approximately 0.24 cpy varies from 35% for the terrestrial part of northern hemisphere to as much as 75% for the OCEAN. (It should be noted that the natural frequency determined from the scalar model of a time series and from different bivariate models where the time series is a component may not be identical.)

The average contribution within that frequency band, without weighting the results in accordance with the areas, amounts to about 60%. On the whole, ENSO undeniably affects the annual surface temperature over the entire globe and over its largest parts and its contribution is significant or even domineering in the vicinity of 0.2 cpy. At lower frequencies, it becomes much smaller.

The innovation sequence contribution to the total variance fluctuates between 30 and 45% (with the LSH result excluded), and it is generated mostly due to the correlation with the innovation sequence $a_{2,t}$ of the NINO time series.

The statistical predictability of AST in the bivariate cases with NINO as the input is better than in the scalar cases but the improvement is always small. In other words, the teleconnection with NINO does not improve the predictability of annual surface temperature averaged over the entire globe and its substantial parts. This result is quite understandable having in mind the small contribution of NINO to the variances

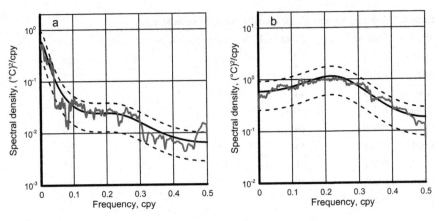

Fig. 8.9 AR (black) and MTM (gray) estimates of GLOBE (**a**) and NINO (**b**) spectra

of all output processes, from GLOBE to LNH, with the baffling exception of southern hemisphere's land. Specifically, the Granger criteria of statistical predictability c_{12} and c_{21} defined with Eqs. (7.29), (7.30) lie roughly within intervals 0.1–0.2 and between 0.03 and 0.05, respectively, and this means that the feedback criterion $s_{12} = c_{12}c_{21}$ never exceeds 0.01 showing the lack of strong influence of feedback upon Granger causality within all teleconnection systems.

8.2.2 Frequency Domain Analysis

The frequency domain analysis constitutes an obligatory part of teleconnection research, and it begins with spectral density estimation. The dynamic range of the spectral density of averaged surface temperature data is always much higher than the range of NINO's spectrum. An example is shown in Fig. 8.9.

The coherence functions that characterize the frequency domain dependence of each time series upon NINO are quite high for all nine systems and their maxima vary between 0.60 (LNH/NINO) and 0.88 (OCEAN/NINO). In particular, the coherence of the GLOBE /NINO system achieves 0.84 at $f = 0.24$ cpy and it stays statistically significant at frequencies from 0.1 to 0.4 cpy (Fig. 8.10a). The high coherence is responsible for the significant—up to 70%—contribution of NINO to the GLOBE spectrum at frequencies between 0.2 and 0.3 cpy (Fig. 8.10b).

The residual (noise) spectrum $n_{11.2}(f)$ plays a dominant role in this and other bivariate time series relating ENSO to spatially averaged annual surface temperature. In the GLOBAL/NINO system, it explains 83% of the GLOBE spectrum $s_{11}(f)$; the remaining 17% are generated by NINO but NINO's contribution within the frequency band from 0.20 cpy to 0.25 cpy to the entire variance of global temperature is negligibly small: just about 3%. The other eight systems behave in the same manner. This result proves that the approach to teleconnection research based upon theory of

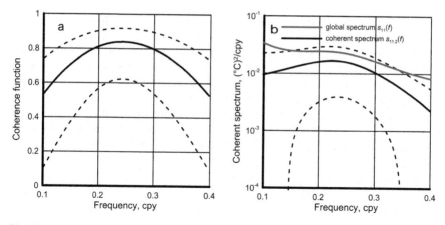

Fig. 8.10 Coherence function (**a**) and coherent (black) and residual (gray) spectra (**b**) in the GLOBE/NINO system

stationary random processes and information theory allows one to detect even very subtle phenomena. In this case, the contribution of ENSO to the total variance of the global temperature is tiny but found to be quite strong within the frequency band related to ENSO.

The part of the GLOBE spectrum coherent to NINO is always statistically significant: within the frequency band from 0.2 cpy to 0.3 cpy, it amounts to almost 70% of the GLOBE spectrum, 75% of the ocean spectrum, and slightly less than 50% of the land spectrum.

The transformation of NINO variations into variations of the global surface temperature is described with the gain factor $g_{12}(f)$. As seen from Fig. 8.11a, the response of the global surface temperature to NINO within the frequency band from

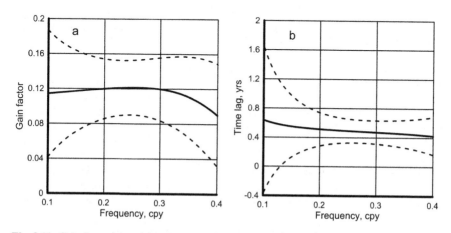

Fig. 8.11 Gain factor (**a**) and time lag (**b**) in the GLOBE/NINO system

0.1 cpy to 0.3 cpy amounts to approximately 0.12 °C per 1 °C change in the NINO temperature.

The physical system GLOBAL/NINO responds to the input process with a delay, which decreases from about 0.6 year at $f = 0.1$ cpy to 0.4 year at $f = 0.4$ cpy. However, the approximate confidence intervals for the estimates are rather wide and allow the gain factor and time lag to be constant over the entire band from 0.1 to 0.4 cpy where the coherence function remains statistically significant. At frequencies below 0.1 cpy, the confidence interval for the gain factor estimate contains zero, which shows that the response of global temperature to NINO is not significant at low frequencies.

This part of the Sect. 8.2.2 presents rather detailed quantitative information about the frequency-dependent properties of teleconnection between ENSO and the annual surface temperature averaged over the globe. Qualitatively, the behavior of the other eight bivariate systems is similar to what has been obtained up to now and it is summed up in Table 8.3 for the frequencies where the coherence function of each system achieves its maximum. As follows from the table, the maximum coherence between NINO and AST can be quite high, especially for the World Ocean and the oceanic southern hemisphere and it occurs at frequencies between 0.23 and 0.26 cpy, that is, close to the natural frequency of ENSO $f_e \approx 0.24$ cpy. The response of temperature to ENSO is smaller for the ocean and higher for the land (up to 0.12 °C against 0.17 °C per 1 °C change in the NINO temperature), while the time lag between the input process and the response to it is higher for the land, especially in the northern hemisphere with its large percentage of terrestrial areas. These results seem to be physically reasonable but their interpretation requires a physical model (or models) of processes in the atmosphere and ocean that would explain ENSO's effects upon the output processes. Besides, the results given in Table 8.3 present statistical estimates and one needs to know respective confidence limits, which have been computed but not shown here in full.

Table 8.3 Frequency domain properties of teleconnections between NINO and annual surface temperature at the frequency of maximum coherence

Output	p	f, cpy	$\max\{\gamma_{12}(f)\}$	$g_{12}(f)$	$\tau_{12}(f)$, year
Globe	2	0.24	0.84	0.12	0.49
NH	2	0.25	0.74	0.12	0.58
SH	2	0.24	0.85	0.12	0.42
Ocean	2	0.25	0.88	0.12	0.42
ONH	2	0.24	0.83	0.11	0.49
OSH	2	0.23	0.78	0.10	0.44
Land	2	0.26	0.71	0.14	0.70
LNH	2	0.26	0.60	0.14	0.81
LSH	2	0.24	0.81	0.17	0.53
Average	2	0.24	0.78	0.13	0.54

The results of time and frequency domain analyses can be formulated in the following manner:

- the ENSO phenomenon represented with the sea surface temperature NINO definitely affects the annual surface temperature in all nine major parts of the globe discussed here.
- the influence of ENSO upon the annual surface temperature is minor because it occurs within the frequency band where the output spectra are much smaller than at lower frequencies.
- the annual surface temperature dependence upon the white noise component is smaller than for NINO: about 30% against approximately 40–50%.
- the coherence functions relating NINO to surface temperature can be as high as 0.88, which makes NINO responsible for up to 77% of spectral density within the frequency band from 0.2 to 0.3 cpy.
- the switch from scalar to bivariate models with NINO as the input does not improve statistical predictability of annual surface temperature.
- a change of 1 °C in NINO causes the response of AST of about 0.1 °C at frequencies between 0.1 and 0.3 cpy.
- it takes from 0.4 year to 0.8 year for the NINO's effect to be seen in the output processes.
- the Granger causality and feedback criteria are small.

It should be remembered that these results have been obtained for the dependence between NINO and time series of spatially averaged surface temperature from which the linear trend has been deleted. If the trend is regarded as a nature-caused part of surface temperature variations, the dependence upon NINO will remain strong within the frequency band from 0.2 cpy to 0.3 cpy but quantitatively the NINO contribution will be negligibly small.

This concludes Example 8.2.

The teleconnection examples given in this part of Chap. 8 demonstrate, among other things, how to study the dependence between time series in general and provide detailed quantitative information about the teleconnections between ENSO and SOI and within the nine bivariate systems AST/NINO.

In the first case (Example 8.1), the relation within the ENSO system represented with SOI and NINO time series is quite strong within a wide frequency band that is responsible for most of the time series variances. However, the presence of the teleconnection does not improve statistical predictability of SOI and NINO in the bivariate case because the high coherence between them occurs due to the dominant contribution of mutually correlated and unpredictable innovation sequence.

In Example 8.2 dedicated to ENSO's influence upon the annual surface temperature over the planet, the high coherence occurs within the frequency band where the spectrum of AST is low. The fact of the teleconnection existence has been established; its quantitative indicators have been obtained but the effect of ENSO upon AST turned out to be small. Both examples prove that the suggested time and frequency domain approach allows one to study teleconnections and obtain reliable results even when the role of teleconnection is very weak.

Hopefully, the main goal of this chapter—to describe the process of teleconnection analysis—has been achieved and the ten cases described here contain both qualitative and quantitative information about each teleconnection.

To the best of this author's knowledge, the only other examples of teleconnection analysis within the framework of theory of random processes and information theory can be found in Privalsky (1988, 2015), Privalsky and Jensen (1995) and, to some extent, in Park and Dusek (2013).

8.3 Example 3. Bivariate Extrapolation of Madden–Julian Oscillation

In this example, the task of extrapolating a bivariate time series within the framework of Kolmogorov–Wiener theory is discussed using as example the Julian–Madden oscillation data. The main goal in this case is to show how to determine the extrapolation error variance in the bivariate case. The solution given here is valid for multivariate time series as well.

The MJO's components RMM1 and RMM2 define the system's spatial position as a function of time. Both time series have been studied in Chap. 6 as samples of scalar stationary processes and shown to be predictable for up to six or seven days at a daily sampling rate. The length of the bivariate time series RMM1/RMM2 is still the same 14,000 days (#4 in Appendix).

The optimal bivariate model $\mathbf{AR}(7)$ for MJO selected by three-order selection criteria is too cumbersome for a methodical task. Therefore, the AR model studied in what follows is the $\mathbf{AR}(4)$ which has been selected by two out of the five OSCs. The differences between the $\mathbf{AR}(4)$ and $\mathbf{AR}(7)$ models are not substantial. The behavior of this process can hardly be regarded as a typical example of teleconnection but the presence of a teleconnection is obvious. As the MJO components carry information about each other, the bivariate approach to their extrapolation may result in better forecasts than in the scalar cases. The MJO process represented with RMM1 and RMM2 data reveals a definite cycling behavior which is illustrated with Fig. 8.12. Obviously, the typical trajectory is close to a circle so if the phase angle of one component is α, the phase angle of the other component will be close to $\pi/2 - \alpha$ (a circular trajectory distorted by noise). Thus, the components do carry information about each other.

This conclusion is also confirmed by the shape of the cross-correlation function between RMM1 and RMM2 (Fig. 8.13a) which shows an interdependent behavior at positive and negative lags from approximately 5–10 days and, borrowing from the later results, by the high coherence between RMM1 and RMM2 at frequencies in the vicinity of 0.02 cpd (Fig. 8.13b). Further bivariate spectral analysis of the time series $\mathbf{x}_t$ at the frequency band where the spectral density is high shows that the real part of frequency response function connecting the time series components (the gain

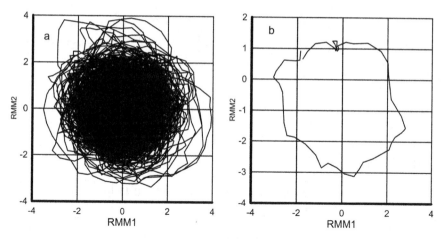

Fig. 8.12 MJO on a plane: 14000 days (**a**) and days 801–850 (**b**)

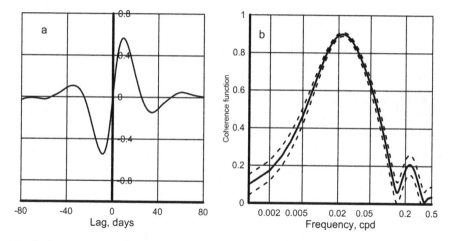

Fig. 8.13 Cross-correlation (**a**) and coherence (**b**) functions between RMM1 and RMM2

factor) can reach 0.95 while the imaginary part (the phase factor) is very close to $\pi/2$.

In this case, the choice of the input and output processes is not important because the MJO components are related to each other more or less deterministically in the phase space while the random part of the process is the varying length of the vector formed by the components. For definiteness, the MJO system is assumed to consist of the input process RMM2 ($x_{2,t}$) and the output RMM1 ($x_{1,t}$). The **AR**(4) model chosen for the time series can be written as

$$x_{1,t} = \varphi_{11}^{(1)} x_{1,t-1} + \varphi_{12}^{(1)} x_{2,t-1} + \cdots + \varphi_{11}^{(4)} x_{1,t-4} + \varphi_{12}^{(4)} x_{2,t-4} + a_{1,t}$$

$$x_{2,t} = \varphi_{21}^{(1)} x_{1,t-1} + \varphi_{22}^{(1)} x_{2,t-1} + \cdots + \varphi_{21}^{(4)} x_{1,t-4} + \varphi_{22}^{(4)} x_{2,t-4} + a_{2,t} \qquad (8.5)$$

or

$$\mathbf{x}_t = \sum_{j=1}^{4} \boldsymbol{\Phi}_j \mathbf{x}_{t-j} + \mathbf{a}_t. \qquad (8.6)$$

The AR coefficients

$$\boldsymbol{\Phi}_1 \approx \begin{bmatrix} 1.53 & \mathbf{0.01} \\ \mathbf{0.01} & 1.46 \end{bmatrix}, \quad \boldsymbol{\Phi}_2 \approx \begin{bmatrix} -0.65 & -0.11 \\ 0.09 & -0.55 \end{bmatrix}$$

$$\boldsymbol{\Phi}_3 \approx \begin{bmatrix} 0.12 & -0.03 \\ 0 & 0.06 \end{bmatrix}, \quad \boldsymbol{\Phi}_4 \approx \begin{bmatrix} -0.03 & 0.07 \\ -0.04 & 0 \end{bmatrix} \qquad (8.7)$$

and the covariance matrix of the innovation sequence $\mathbf{a}_t = [a_{1,t}, a_{2,t}]'$ is

$$\mathbf{P_a} \approx \begin{bmatrix} 0.030 & 0 \\ 0 & 0.029 \end{bmatrix}. \qquad (8.8)$$

The coefficients shown in bold are statistically insignificant.

The results of analysis given by Eqs. (8.7) and (8.8) reveal the following features of the time series $\mathbf{x}_t$:

- each component of the time series depends mostly upon its behavior in the past (all but one diagonal coefficients are statistically significant and relatively large).
- this is a closed feedback loop system but the influence of PMM1 and PMM2 upon each other is minor (smaller or even statistically insignificant nondiagonal elements).
- the innovation sequences are not correlated to each other.

Consider now the task of extrapolating MJO as a bivariate time series. The trajectory of PMM1 and PMM2 predicted at the lead time τ days is

$$\hat{\mathbf{x}}_{t+\tau} = \sum_{j=1}^{4} \boldsymbol{\Phi}_j \hat{\mathbf{x}}_{t+\tau-j} \qquad (8.9)$$

and the terms $\hat{\mathbf{x}}_{t+\tau-j}$ coincide with the observed values if $\tau - j < 0$. The bivariate time series given with Eqs. (8.5) or (8.6) with the matrix AR coefficients (8.7) is stationary so that the predicted trajectories of the components will be tending to zero (the mean values) similar to the behavior of the scalar component RMM1 in Chap. 6. But the more important issue is whether the prediction error variance in the bivariate case is smaller than the error variance of the scalar extrapolation discussed in Chap. 6

for the component RMM1. The following comparison will be done for RMM1 but the results are practically the same for RMM2.

The RMM1 variance $\sigma_1^2 \approx 1.01$ while the innovation sequence variance $P_{11} \approx 0.030$, that is, the RMM1 component has high statistical predictability at $\tau = 1$ day. In the scalar case, the extrapolation error variance is a few percent higher but the task here is to compare the extrapolation error variance for the scalar and bivariate cases at lead times longer than one day. For this purpose, it is necessary to determine the matrix coefficients of the operator $\boldsymbol{\Psi}(B) = \boldsymbol{\Phi}^{-1}(B)$, similar to what has been done in Chap. 6 for the scalar coefficients [see Eqs. (8.8–8.10)]. The coefficients of the operator $\boldsymbol{\Phi}(B)$ are given with Eq. (8.7) so that the matrix coefficients $\boldsymbol{\Psi}_j$ are found by equating coefficients at equal powers of B in operators $\boldsymbol{\Phi}^{-1}(B)$ and $\boldsymbol{\Psi}(B)$:

$$\boldsymbol{\Psi}(B) = \boldsymbol{\Phi}^{-1}(B) = \sum_{j=0}^{\infty} \boldsymbol{\Psi}_j B^j \tag{8.10}$$

The vector extrapolation error $\boldsymbol{\varepsilon}_{t+\tau} = \mathbf{x}_{t+\tau} - \hat{\mathbf{x}}_{t+\tau}$ can be presented as

$$\boldsymbol{\varepsilon}_{t+\tau} = \sum_{j=0}^{\tau-1} \boldsymbol{\Psi}_j \mathbf{a}_{t+\tau-j} \tag{8.11}$$

so that the error variance at lead time τ is

$$\boldsymbol{\Sigma}^2(\tau) = E[\boldsymbol{\varepsilon}_{t+\tau}\boldsymbol{\varepsilon}'_{t+\tau}] = \sum_{j=0}^{\tau-1} \boldsymbol{\Psi}_j \mathbf{P_a} \boldsymbol{\Psi}'_j, \tag{8.12}$$

where E is the operator of mathematical expectation and the strike means matrix or vector transpose. The last equation is a multivariate extension of Eq. (6.11) for extrapolation of scalar time series, and it can be found in Reinsel (2003) and in Box et al. (2015). The prediction error variances for the components of an M-variate time series are the diagonal elements of the matrix $\boldsymbol{\Sigma}^2(\tau)$.

Let $\rho(\tau) = \sigma_\varepsilon(\tau)/\sigma_x$ given with Eq. (6.13) and $P_{11}(\tau) = \Sigma_{11}(\tau)/\sigma_1$ be the relative predictability criteria for PMM1 as a scalar time series and as a component of the bivariate time series PMM1/PMM2, respectively. Here, $\Sigma_{11}(\tau)$ is the square root of the first diagonal element of the matrix $\boldsymbol{\Sigma}^2(\tau)$. The bivariate prediction of PMM1 is more reliable than in the scalar case but the difference is not large: the bivariate criterion is smaller than the scalar one by not more than 10–12% (Fig. 8.14). With the lead time growing, the two error variances tend to the common limit—the variance σ_1^2 of the component RMM1. As the standard deviation σ_1 is very close to 1, the curves in Fig. 8.14 actually show the standard deviations of forecast errors of RMM1 as a function of lead time. The extrapolation error properties for the component RMM2 are practically the same.

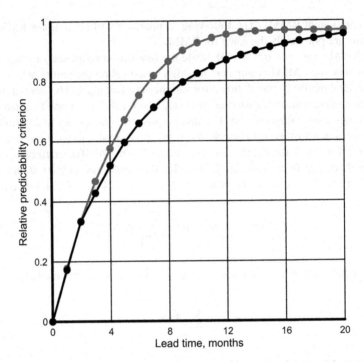

Fig. 8.14 Relative predictability criteria for RMM1 as a scalar time series (gray) and as a component of the bivariate time series (black)

In this case, the predictability turned out to be just slightly higher in the bivariate case which probably happened because of the weak feedback between the components. The Granger causality criteria c_{12} and c_{21} equal approximately 0.04 and 0.06, respectfully, which means the lack of causality and practically zero feedback criterion $s_{12} = c_{12}c_{21}$.

If one introduces an extended criterion of Granger causality as a function of lead time, that is,

$$c_{12}(\tau) = [1 - v_{12}(\tau)/v_{11}(\tau)],$$

and respectfully for c_{21}, the causality and feedback criteria will differ but the changes will not be significant.

Summing up, a multivariate approach to forecasting of stationary time series does not necessarily lead to perceptible improvement of statistical predictability unless the time series components do not show a steady time lag.

Appendix

1. http://www.bom.gov.au/climate/current/soi2.shtml
2. https://climexp.knmi.nl/data/ihadisst1_nino3.4a.dat
3. https://crudata.uea.ac.uk/cru/data/temperature/HadCRUT4-gl.dat
4. http://www.bom.gov.au/climate/mjo/graphics/rmm.74toRealtime.txt.

References

Box G, Jenkins M, Reinsel G, Liung G (2015) Time series analysis. Forecasting and control, 5th edn. Wiley, Hoboken

Gelfand I, Yaglom A (1957) Calculation of the amount of information about a random function contained in another such function, Uspekhi Matematicheskikh Nauk, 12:3–52, English translation: American Mathematical Society Translation Series 2(12):199–246, 1959

He S, Yang S, Lu M, Li Z (2018) Afro-Eurasian intermediate-frequency teleconnection and modulation by ENSO. J Climate 31:8121–8139

Park J, Dusek G (2013) ENSO components of the Atlantic multidecadal oscillation and their relation to North Atlantic interannual coastal sea level anomalies. Ocean Sci 9:535–543

Privalsky V (1988) Stochastic models and spectra of interannual variability of mean annual sea surface temperature in the North Atlantic. Dyn Atmos Oceans 12:1–18

Privalsky V (2015) On studying relations between time series in climatology. Earth Syst Dyn 6:389–398

Privalsky V, Jensen D (1995) Assessment of the influence of ENSO on annual global air temperature. Dyn Atmos Oceans 22:161–178

Privalsky V, Muzylev S (2013) An experimental stochastic model of the El Niño—Southern oscillation system at climatic time scales. Int J Geosci https://doi.org/10.13189/ujg.2013.010104

Privalsky V, Yushkov V (2018) Getting it right matters: climate spectra and their estimation. Pure Appl Geophys 175:3085–3096

Reinsel G (2003) Elements of multivariate time series analysis, 3rd edn. Springer, New York

Yeh S-W, Cai W, Min S-K et al (2018) Enso atmospheric teleconnections and their response to greenhouse gas forcing. Geophys, Rev. https://doi.org/10.1002/2017RG000568

Yuang C, Wang D (2019) Interdecadal variations in ENSO impacts on the autumn precipitation in the Eastern China. Int J Climatol. https://doi.org/10.1002/joc.6156

Chapter 9
Reconstruction of Time Series

Abstract Time series reconstruction is a common task in Earth and solar sciences, especially in climatology. A short target series is reconstructed into the past using its relation to a longer proxy (or proxies) time series known over a short interval of simultaneous observations. The traditional regression method of reconstruction is mathematically incorrect. The author's method based upon the theory of random processes requires autoregressive modeling of the initial target/proxy time series in time and frequency domains; the modeling results are used to reconstruct the target. The method is tested here through the use of a simulated bivariate time series of 1500 time units. The proxy/target relation obtained for the last 150 values of the time series is applied for reconstructing the target component over the previous 1350 time units. The time and frequency domain properties of the reconstructed time series are compared with the respective properties of the true target over the same long interval. Results of reconstruction are shown to agree with the true properties of the target. The share of the variance of the target time series reconstructed with the method amounts to 70% while the traditional correlation/regression method cannot restore more than 30% of target's variance.

9.1 Introduction

The task of time series reconstruction exists in different branches of Earth sciences and in solar physics. The term "reconstruction" in this context means generating a previously unavailable target time series, which is known only partially over a short recent time interval and which needs to be extended, or reconstructed, over some time interval in the past. This task is solved by using the available relatively short target time series together with one or several proxy time series known over the same interval as the target and extending far enough into the past. A dependence between the target and proxy series is established for the common time interval and then used to calculate (reconstruct) the target time series over the interval in the past for which the proxy is (are) known. A typical example would be a reconstruction of annual surface temperature or precipitation as the target using the dependence between the surface temperature and a proxy in the form of a tree ring widths time series, which

© Springer Nature Switzerland AG 2021
V. Privalsky, *Time Series Analysis in Climatology and Related Sciences*,
Progress in Geophysics, https://doi.org/10.1007/978-3-030-58055-1_9

is known simultaneously with the target for a recent time interval and also extends for several hundred years into the past.

Time series reconstruction is a common task in climatology and paleoclimatology, in solar physics, hydrology, oceanography, atmospheric science, hydrometeorology, and hydrogeology, and in other Earth sciences and at different sampling rates, from hours and days to years or from months, years, and decades to centuries and millennia. In solar physics, it is often reconstruction of time series of solar irradiance, total or spectral (i.e., within specific wavelength bands). Numerous examples can be found in many if not all popular Earth and solar science journals including *Journal of Climate* (e.g., Schweiger et al. 2019), *International Journal of Climatology* (Oyunmunkh et al. 2019; Peng et al. 2020), Climate of the Past (Ahmad et al. 2020), *Geophysical Research Letters* (Bryan et al. 2019; Zaw et al. 2020), *Climate Dynamics* (Adloff et al. 2018; Seftigen et al. 2020), and *The Holocene* (Lepley et al. 2020). *The Journal of Geophysical Research* has published articles on time series reconstruction in at least five of its seven sections: space physics (Flynn et al. 2018), oceans (Taniguchi et al. 2018), atmospheres (Wang et al. 2019), Earth surface (Noël et al. 2018), biogeosciences (Maxwell et al. 2018). The number of articles published on the subject for the last several decades in leading climate and Earth and solar science journals probably exceeds five hundred. The same approach can be found in monographs on teleconnections and solar influence upon climate (e.g., Roy 2018).

The reconstruction studies can have different purposes and different types of data but the most popular subject is the reconstruction of time series into the past using climate-related proxies with the goal to restore past climates. As seen from the latest report of the International Panel on Climate Change (IPCC 2013), time series of reconstructed geophysical variables at climatic time scales are essential for generating "projections" of the global warming issued by the IPCC. In what follows, we will describe the traditional method of time series reconstruction and the method suggested by the author of this book.

9.2   Methods of Reconstruction

The task of time series reconstruction is not known in the theory of random processes and theoretical time series analysis; the tool that is traditionally used for it in Earth and solar sciences belongs to the classical mathematical statistics: it is the linear regression equation relating the target to a proxy (or proxies). If the correlation coefficient between the target and proxy is sufficiently high, the reconstruction is believed to restore a sizable share of the target's variance. This is the approach which is specifically not recommended for studying relations between time series (Privalsky and Jensen 1995; Thomson and Emery 2014, p. 433) but it has a long history and continues to play the domineering role in climatology, solar science, and in other Earth sciences. Therefore, the subject is important enough to be discussed in the light of results relevant for the reconstruction task which have appeared in information theory and time series analysis more than half a century ago.

The current traditional regression approach was first suggested by the American astronomer and archeologist Andrew Ellicott Douglass (1867–1962) who is justly recognized as the father of dendrochronology and climate reconstructions in general. His ideas and results are presented in a series of treatises and articles published between 1909 and 1936. Professor Douglass was the first to discover that "the [tree] rings are likely to form a measure of the precipitation" (Douglass 1909, p. 226), that "the mean growth of the tree is proportional to accumulated rain," and that respective equations were able to explain up to 80% of the dependence between two variables (Douglass 1914, p. 327). He also understood that the tree growth is affected by temperature (Douglass 1928, p. 102). The method of climate reconstruction based upon A. Douglass' ideas is briefly described below.

9.2.1 Traditional Correlation/Regression Reconstruction (CRR)

Let $\mathbf{y} = [y_{N+1}, \ldots, y_{N+M}]'$ and $\mathbf{x} = [x_{N+1}, \ldots, x_{N+M}]'$ be simultaneous observations of zero mean random vectors $\mathbf{y}$ and $\mathbf{x}$, which are called the target and the proxy, respectively. The proxy values x_n are also known over the interval from $n = 1$ through $n = N$. If the cross-correlation coefficient r_{xy} between the target and proxy estimated for the interval of simultaneous observations from $N + 1$ through $N + M$ is high, the relation between the target and proxy can be expressed in the form of a linear regression equation $\mathbf{y} = b\mathbf{x}$, where b is the regression coefficient. Assuming that the coefficients r_{xy} and b estimated with data from $N + 1$ through $N + M$ have been the same for the time interval between $n = 1$ and $n = N$, the past values of the target can be reconstructed as

$$\hat{y}_n = bx_n, \ n = 1, \ldots, N. \tag{9.1}$$

The error of this estimate is

$$\eta_n = y_n - \hat{y}_n, \ n = 1, \ldots, N. \tag{9.2}$$

The true values of the target in the past are not known but the error variance is defined as

$$\sigma_\eta^2 = (1 - r_{xy}^2)\sigma_y^2. \tag{9.3}$$

This is the method first introduced by Douglass (1914). According to the recent fundamental review of temperature reconstructions by Christiansen and Ljungkvist (2017), the methods of reconstruction used in Earth and solar sciences are practically always based upon linear regression. This means that the basics of the reconstruction method have not changed since more than a century ago. The terminology in the latest climate reconstruction literature can be different: time series may become variables and the regression equation may be called transfer function (Liu and Li 2019) but it is the same method: reconstruction through a linear regression equation.

However, the author of the method A. Douglass was not totally satisfied with his own approach to reconstruction; he believed that "the methods of computing rainfall from tree growth must be still further perfected" (Douglass 1914, p. 335). The author of the CRR method saw the following areas that required improvements. Firstly, A. Douglass was dissatisfied with the cross-correlation coefficient as a measure of dependence between time-dependent random data: "the similarity between two curves is … only partially expressed by a correlation coefficient" (Douglass 1936, p. 29). He and his colleagues tried to construct "an index of similarity that is more satisfactory than the correlation coefficient" (*ibid.*) but were not successful. As shown in Chap. 7 of this book, the "index" they were looking for is the coherence function, which belongs to the frequency domain analysis of multivariate time series unknown at that time (along with the term "time series"). A. Douglass understood the inadequacy of the cross-correlation coefficient as a measure of interdependence between time series, that is, between random functions of time, long before the issue has been resolved in information theory by I. Gelfand and A. Yaglom in their classical work (Gelfand and Yaglom 1957).

Secondly, A. Douglass saw the necessity to assume that "the ring growth in one year was built up by contributions from the current year and previous years …" (Douglass 1919, p. 68). This actually means a necessity to take into account the dependence of tree growth upon itself and upon rainfall for several previous years—a way to create a bivariate autoregressive system relating rainfall to tree growth. It took half a century for the concept of multivariate parametric models in the form of stochastic difference equations to become a practical tool for solving the mathematical problems raised by A. Douglass (Box and Jenkins 1970).

Thus, A. Douglass wanted to change the random variables in Eq. (9.1) to random functions of time ("curves"), which depend upon their behavior in the past and to find a mathematical tool that could be used to describe relations between time-dependent random functions.

The problem with the almost universal application of regression methods is caused by the simultaneous use of the term "time series" and the cross-correlation coefficient between the target and proxy. As the cross-correlation coefficient does not characterize relations between time series, it makes the regression methods mathematically improper in all cases but one—when both target and proxy time series (scalar or multivariate) present white noise, that is, a sequence of identically distributed and mutually independent (or uncorrelated, in the non-Gaussian case) random variables which do not depend upon time by definition. Actually, the dependence between time series is described with the cross-correlation function in the time domain and the coherence function in the frequency domain. It can also be seen explicitly in the time domain multivariate parametric models (autoregressive or mixed autoregressive and moving average).

One of the consequences of the use of cross-correlation coefficient for time series reconstruction is the incorrect determination of statistical reliability of that coefficient. Being a single value of the cross-correlation function, it depends on the entire correlation matrix $\mathbf{r}(k)$ of the time series $\mathbf{x}_t = [x_{1,t}, x_{2,t}]'$. In the correlation matrix given by Eq. (7.11), the estimated cross-correlation coefficient $r_{12}(0)$ between time series $x_{1,t}$ and $x_{2,t}$ cannot be treated as a separate quantity—its reliability must be determined along with the determination of reliability of the entire correlation matrix which, by itself, is by no means an easy task (e.g., Bendat and Piersol 2010). The traditional way to determine the reliability of the cross-correlation coefficient between time series in the manner that it is applied in mathematical statistics for time-invariant random variables is mathematically wrong.

Moreover, the use of even the cross-correlation function for reconstruction purposes is problematic. A simple example is the bivariate time series of total solar irradiance of the Earth (TSI) and the sunspot numbers (SSN). The estimated cross-correlation function $r_{12}(k)$ between these time series shown in Fig. 9.1 contains 13 values exceeding the cross-correlation coefficient $r_{12}(0)$. Obviously, reconstructing the target time series of TSI through the proxy time series SSN using the cross-correlation function $r_{12}(k)$ will be difficult or even impossible. Similar situations may occur in Earth sciences.

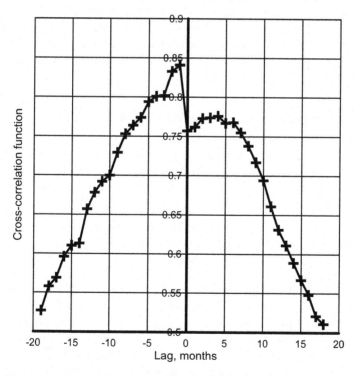

Fig. 9.1 Cross-correlation function between TSI and SSN

Another shortcoming of the regression method of time series reconstruction is the lack of frequency domain information caused by the absence of both time and frequency domains for random variables. The spectral density and the spectral matrix are the most essential characteristics of scalar or multivariate time series, and they provide the information about the target and proxy time series and about their inter-dependence that is required for mathematically correct reconstruction. The information can also be used for evaluating the quality of proxies and for determining the contribution of noise to the proxy and target time series (see Chap. 12).

The deficiencies of the traditional CRR approach are corrected in the method of autoregressive reconstruction (ARR) first introduced in Privalsky (2018).

The CRR method has two other drawbacks not mentioned by A. Douglass in his pioneering research. Firstly, the regression Eq. (9.1) does not include a time delay between the target and proxy that exists in any physically realizable system. Generally, this feature is inbuilt into any type of the regression equation (linear or nonlinear, univariate or multivariate). Secondly, when the relation between time series is expressed with Eq. (9.1), the spectral densities of the target and proxy time series have identical shapes, which, of course, cannot be true for all time series. However, these drawbacks cannot be blamed upon the regression equation for the simple reason that this equation is applicable only to time-invariant random vectors which do not have correlation functions and spectral densities.

Studying relations between time series requires mathematical tools lying within the framework of theory of random processes and information theory. Some basics of bivariate time series analysis are given in Chap. 7 along with several references to mathematical literature written for readers working in applied sciences.

9.2.2 Autoregressive Reconstruction Method (ARR)

A description of the method has been given in Privalsky (2018). Having in mind the importance of reconstruction task in many disciplines, especially in climatology, as well as the numerous publications based upon the outdated CRR method, the suggested ARR method will be described here in more detail.

1. Let $\mathbf{x}_t = [x_{1,t}, x_{2,t}]'$ be a zero mean stationary time series whose components $x_{1,t}$ and $x_{2,t}$ are known for the time intervals $t = T_1 + 1, \ldots, T_2$ and $t = 1, \ldots, T_1, \ldots, T_2$, respectively. (Here, time $t = n\Delta t$ and, in what follows, $\Delta t = 1$ year, month, day, etc.) The goal is to reconstruct the past values of the target time series $x_{1,t}$ for values of t from $t = 1$ through $t = T_1$. It is achieved here by building a bivariate AR model of the time series $\mathbf{x}_t$ for $t = T_1 + 1, \ldots, T_2$ and using it for reconstructing the target $x_{1,t}$ over the interval $t = 1, \ldots, T_1$ for which the values of the proxy $x_{2,t}$ are known. The stages of this analysis are described below in accordance with the theoretical basics given in Chap. 7.

2. Calculate an estimate of the correlation matrix $\mathbf{r}(k)$ relating $x_{1,t}$ to $x_{2,t}$ over the interval of simultaneous observations from $t = T_1 + 1$ through $t = T_2$. The

length of this bivariate time series is $T_2 - T_1$ and the recommended absolute values of the lag k should not exceed $(T_2 - T_1)/10$. Of most interest here is the cross-correlation function $r_{12}(k) = r_{21}(-k)$, which gives one an idea of how simple or difficult the reconstruction task will be in the case at hand. Otherwise, the method does not require the knowledge of the cross-correlation function.

3. Select an optimal **AR** model for the bivariate time series. According to Chap. 7, it means building a set of **AR** models for orders p from $p = 1$ through $p = p_{max}$. The maximum order p_{max} should be chosen in such a way that the number of AR coefficients which need to be estimated would not exceed approximately 0.1 of time series length. Otherwise, the estimates could be statistically unreliable. The optimal order of the model must be chosen with the help of order selection criteria given in Chap. 7 for multivariate AR models (also see Reinsel 2003; Box et al. 2015; Shumway and Stoffer 2017). The use of the criteria allows one to avoid selecting an improper order; if the order is too low or exceedingly high, the order selection criteria will probably prevent an erroneous choice.

4. The information about the selected **AR** model should include estimates of the matrix **AR** coefficients (with their error variances) and the covariance matrix of the bivariate innovation sequence along with estimates of respective error variances. It can be obtained, in particular, with the R package VAR given in Shumway and Stoffer (2017) and in MATLAB.

The time domain bivariate **AR** model of order p has the form

$$x_{1,t} = \varphi_{11}^{(1)} x_{1,t-1} + \varphi_{12}^{(1)} x_{2,t-1} + \cdots + \varphi_{11}^{(p)} x_{1,t-p} + \varphi_{12}^{(p)} x_{2,t-p} + a_{1,t}$$
$$x_{2,t} = \varphi_{21}^{(1)} x_{1,t-1} + \varphi_{22}^{(1)} x_{2,t-1} + \cdots + \varphi_{21}^{(p)} x_{1,t-p} + \varphi_{22}^{(p)} x_{2,t-p} + a_{2,t}. \quad (9.4)$$

The covariance matrix of the bivariate innovation sequence $\mathbf{a}_t = [a_{1,t}, a_{2,t}]'$ is

$$\mathbf{P_a} = \begin{bmatrix} P_{11} & P_{12} \\ P_{21} & P_{22} \end{bmatrix}. \quad (9.5)$$

The correlation coefficient between the components of the innovation sequence is $\rho_{12} = P_{12}/\sqrt{P_{11}P_{22}}$. It may play an important role in time series reconstruction.

5. The reconstructed values of the target time series $x_{1,t}$ are then found as

$$\hat{x}_{1,t} = \varphi_{11}^{(1)} \hat{x}_{1,t-1} + \varphi_{12}^{(1)} x_{2,t-1} + \cdots + \varphi_{11}^{(p)} \hat{x}_{1,t-p} + \varphi_{12}^{(p)} x_{2,t-p}. \quad (9.6)$$

The first reconstructed value can be obtained for $t = p + 1$ because the initial values of $\hat{x}_{1,t-p}$ for $t \leq p$ should be zeroes (i.e., the mean value of the time series). This stage concludes the reconstruction of the time series $x_{1,t}$ in the time domain. The reconstruction cannot be regarded as complete unless the properties of the observed and reconstructed time series are studied in the frequency domain. This includes the following steps.

6. Use the time domain model (9.4), (9.5) to calculate the spectral matrix

$$\mathbf{s}(f) = \begin{bmatrix} s_{11}(f) \, s_{12}f) \\ s_{21}(f) \, s_{22}f) \end{bmatrix}, \quad 0 \le f \le f_N, \tag{9.7}$$

and other frequency-dependent functions, that is, the coherence function, coherent spectrum, and gain and phase factors. Analyze their behavior and compare the spectral density $\hat{s}_{11}(f)$ of the reconstructed time series $\hat{x}_{1,t}$ with the spectrum $s_{11}(f)$ of the time series $x_{1,t}$ estimated for the time interval $t = T_1 + 1, \ldots, T_2$. The reconstruction can be regarded as acceptable in this respect if the two spectra have sufficiently similar shapes and if the difference $s_{11}(f) - \hat{s}_{11}(f)$ does not contain features that cannot be explained from mathematical and physical considerations.

To make sure that reconstruction is correct, one needs to verify the coherence function between the reconstructed time series $\hat{x}_{1,t}$ and the proxy $x_{2,t}$; it has to be unity at all frequencies because the reconstructed time series is obtained as a linear transformation of the proxy. The degree of linear dependence between the true and reconstructed output time series $x_{1,t}$ and $\hat{x}_{1,t}$ is given by the coherence function estimate $\hat{\gamma}_{\hat{1}1}(f) = |\hat{s}_{\hat{1}1}(f)|/[\hat{s}_{11}(f)s_{11}(f)]^{1/2}$. Here $\hat{s}_{\hat{1}1}(f)$ is the cross-spectrum between the reconstructed and observed target time series while $\hat{s}_{11}(f)$ and $s_{11}(f)$ are respective spectra.

The ARR method treats time series in a mathematically proper way, and it does not require any mathematical tools that have not been known in time series analysis and information theory since many decades ago. The method can be generalized for the multivariate case but one has to remember that the number of parameters that need to be estimated for an M-variate time series is $M^2 p$. If $M = 4$ (three proxies) and the AR order $p = 3$, one will have to estimate 48 autoregressive coefficients. This means that the initial "short" four-variate time series should be about 500 time units long in order to obtain more or less reliable estimates in time and frequency domains. As for the software, the author is not aware of any free or commercial packages including both time and frequency domain analysis of multivariate autoregressive sequences.

9.3 Verification of the Autoregressive Reconstruction Method

The example of time series reconstruction in Privalsky (2018) was dedicated to the estimation of past monthly values of total solar irradiance (TSI) using sunspot numbers (SSN) as a proxy time series. The TSI time series which has been reconstructed in that example was known only over a relatively short time interval of instrumental observations (February 1978–April 2018) so that it can be justly stated that the reconstructed TSI data have not been independently verified. Besides, the CRR method has a long history—over a century—and its use in climatology, solar physics, and other sciences specifically for time series reconstruction has been challenged, leaving aside the recent publication in 2018, with just one practical example of teleconnection analysis (Privalsky and Jensen 1995). It seems that the sole important opponent of the method was its author A. Douglass, who made the above-given critical comments showing his concerns about the CRR method.

The way out of this impasse is to show the advantages of the ARR method through analysis of a bivariate time series in which both components—the proxy and the target—are known for the entire time span from $t = 1$ through $t = T_2$. This would allow one to compare the reconstructed time series with the true past values. The goal of the example given below is two-pronged: to demonstrate the process of autoregressive reconstruction and to verify the method through using simulated data.

9.3.1   Reconstruction Example: A Climatic Type Process

Consider an example when the stationary bivariate time series used for the reconstruction of one of its components behaves in a manner typical for climate variations and probably for other geophysical data. Let the time series $\mathbf{x}_t = [x_{1,t}, x_{2,t}]', t = 1, \ldots T$ be a sample record of a Markov, that is, an AR(1), stationary discreet random process containing an additional contribution from a stationary white noise:

$$x_{1,t} = \varphi x_{1,t-1} + x_{2,t-1} + a_{1,t}$$
$$x_{2,t} = a_{2,t}. \tag{9.8}$$

The Gaussian innovation sequences $a_{1,t}$ and $a_{2,t}$ are mutually independent and the sampling interval is $\Delta t = 1$.

Such phenomenon could be a slightly simplified version of time series common in Earth sciences when the target time series $x_{1,t}$ presents, for example, river streamflow or surface temperature, while the proxy $x_{2,t}$ can be a time series of tree ring widths or thicknesses of sediment layers. The autoregressive coefficient $\varphi = 0.75$, the variance of the innovation sequence $a_{1,t}$ is $P_{11} = 0.5$, the variance of the proxy component $x_{2,t}$ is $\sigma_2^2 = 1$. These are the true values of statistical parameters characterizing the time series $\mathbf{x}_t$. Obviously, the bivariate system expressed with Eq. (9.8) has no feedback which is quite reasonable because one should not expect that the proxy time series such as thickness of tree rings can be affected by streamflow of an individual river. The true variance of the target component $x_{1,t}$ corresponding to these statistics is $(P_{11} + \sigma_2^2)/(1 - \varphi^2) \approx 3.43$.

Both $x_{1,t}$ and $x_{2,t}$ are known for the entire time interval of length $1500\Delta t$. The target $x_{1,t}$ will be reconstructed using the last 150 observations of the bivariate target and proxy time series. The past values of $x_{1,t}$ for the first 1350 time intervals will be reconstructed in accordance with Eq. (9.6), and the reconstructed time series $\hat{x}_{1,t}$ will be compared with the true values of $x_{1,t}$ over that time interval. The target and proxy time series are shown in Fig. 9.2. The initial larger part of the time series $x_{1,t}, t = 1, \ldots, 1350$ will be used only for an independent verification of reconstruction results.

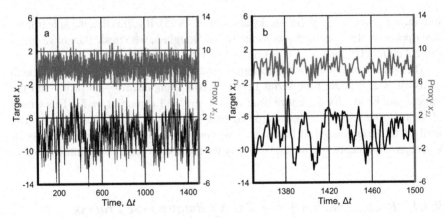

Fig. 9.2 Entire time series x_t (**a**) and its 150 last values (**b**)

The model obtained by analyzing the bivariate time series over the interval $t = 1351, \ldots, 1500$ turned out to be an **AR**(1) sequence

$$x_{1,t} \approx 0.77 x_{1,t-1} + 1.04 x_{2,t-1} + a_{1,t}$$
$$x_{2,t} \approx a_{2,t}. \tag{9.9}$$

This model has been selected by all order selection criteria used in the software. Other parameters of the time series (9.9) are $\sigma_1^2 \approx 3.25$, $P_{11} \approx 0.48$, and $\sigma_2^2 \approx 0.93$. The true values are $3.43, 0.50$, and 1.00, respectively. Thus, the estimates of variances are lower than the true values while the coefficient φ is slightly overestimated. The differences from the true values do not exceed 7%.

The cross-correlation function between the input and output time series (Fig. 9.3) shows a sharp peak at $k = -1$ slightly exceeding 0.5; this behavior allows one to assume that the component time series show at least some interdependence. The true value of the cross-correlation $r_{12}(-1)$ can be obtained by multiplying the equation $x_{1,t} - \varphi x_{1,t-1} + x_{2,t-1} = a_{1,t}$ by $x_{2,t-1} = a_{2,t-1}$ and finding the mathematical expectation: $r_{12}(-1) = \sigma_2/\sigma_1 \approx 0.54$. The estimated cross-correlation $r_{12}(-1)$ for $N = 150$ is 0.51.

According to Eqs. (9.6) and (9.9), the reconstructed time series is

$$\hat{x}_{1,t} = \hat{\varphi} \hat{x}_{1,t-1} + x_{2,t-1}, t = 2, \ldots, 1350, \tag{9.10}$$

where $\hat{\varphi}$ is the estimate of autoregressive parameter φ obtained from the analysis of the time series x_t for the values of t from 1351 through 1500. The initial value coincides with the mean value of $x_{1,t}$, that is, $\hat{x}_{1,1} = 0$. The quality of reconstruction seems to be adequate (Fig. 9.4). Quantitatively, the variance $\hat{\sigma}_x^2$ of the reconstructed time series $\hat{x}_{1,t}$ is approximately 2.50 while the variance σ_x^2 of the true time series $x_{1,t}, t = 2, \ldots, 1350$ is 3.50.

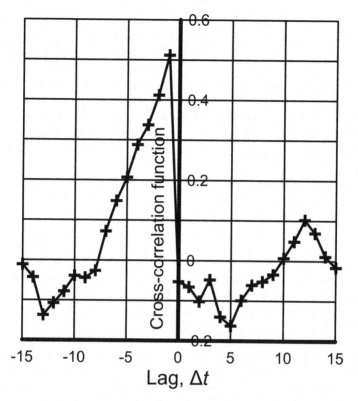

Fig. 9.3 Cross-correlation function between the target and proxy time series ($N = 150$)

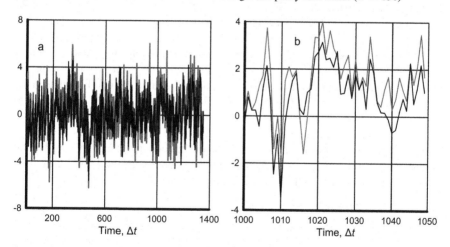

Fig. 9.4 True (gray) and reconstructed (black) target time series $x_{1,t}$ and $\hat{x}_{1,t}$, $t = 2, \ldots, 1350$

Thus, the reconstruction procedure has resulted in rebuilding 71% of the true time series variance. This is substantial and the results of reconstruction should be regarded as satisfactory in that respect. The traditional CRR approach would have reconstructed practically nothing if the cross-correlation coefficient $r_{12}(0)$ were used and about 26% of the target variance if it were the maximum of the cross-correlation function $r_{12}(-1) \approx 0.51$.

According to the above-given results, the reconstruction error variance is close to 1 while, according to Eq. (9.8), the part of $x_{1,t}$ which cannot be reconstructed is the innovation sequence $a_{1,t}$ whose true variance is only 0.50. This difference should be explained.

The reconstruction errors $\varepsilon_{1,t}$ are found as the difference between Eqs. (9.8) and (9.10):

$$\varepsilon_t = \varphi x_{1,t-1} - \hat{\varphi}\hat{x}_{1,t-1} + a_{1,t}, \, t = 2, \ldots, 1350. \tag{9.11}$$

This means that the reconstruction error includes two components which occur due to the difference in statistics of the target time series determined on the basis of the short and long parts of $x_{1,t}$ (from 1351 through 1500 and from 1 through 1350):

- the error in the estimated variance of innovation sequence $a_{1,t}$,
- the error in the value of estimated autoregression parameter(s).

These errors lead to an error in the variance of the initial target time series and to an error in the variance of the reconstructed time series (Table 9.1). Ideally, the variance should have been 0.5 and actually it turned to be twice larger due to the sampling variability of parameter estimates. Such sampling errors occur in all cases of time series reconstruction. Still, the efficiency of reconstruction with the ARR method is much better than what could have been obtained with the traditional CRR method.

As follows from Eq. (9.11), the time series of reconstruction error is not necessarily a white noise sequence because it contains a difference between two time series. In our case, it is the AR(1) model with approximately the same AR coefficient as in the true time series and the variance $\sigma_\varepsilon^2 \approx 1.15$ while the variance $\hat{\sigma}_x^2 \approx 2.50$. The sum $\sigma_\varepsilon^2 + \hat{\sigma}_x^2$ differs from the true variance σ_x^2 by just several percent. This analysis shows that the notion of reconstruction error as a white noise is not necessarily correct.

Thus, the reconstruction errors seem to be understood and it can be said that the time domain properties of the reconstructed data agree with the known true time series over the reconstruction time interval.

Table 9.1 Reconstruction errors

Statistics	True	Short time series (150)	Long time series (1350)
φ	0.750	0.767	0.753
P_{11}	0.500	0.478	0.472
σ_x^2	3.429	3.251	3.496

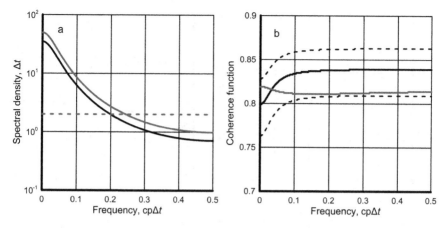

Fig. 9.5 a True (gray) and reconstructed (black) spectra $s_{11}(f)$ and $\hat{s}_{11}(f)$; **b** the true (gray) and reconstruction (black) coherence functions $\gamma_{12}(f)$ and $\hat{\gamma}_{12}(f)$

The next step should be a comparison of frequency domain properties of the actual target $x_{1,t}$ with the reconstructed target time series $\hat{x}_{1,t}, t = 2, \ldots, 1350$. If reconstruction is done correctly, the spectral density $\hat{s}_{11}(f)$ of the reconstructed data ("reconstructed spectrum") and the true target spectrum $s_{11}(f)$ should be similar. As seen from Fig. 9.5a, this condition is definitely satisfied. The spectrum $\hat{s}_{11}(f)$ lies below the spectrum $s_{11}(f)$ because the reconstructed time series has a smaller variance, in particular, due to the exclusion of the innovations sequence $a_{1,t}$. The dashed gray line shows the spectrum of the proxy.

As the reconstructed target time series $\hat{x}_{1,t}$ does not contain any contribution from the white noise sequence $a_{1,t}$, the coherence $\hat{\gamma}_{12}(f)$ between the true target time series $x_{1,t}$ and the reconstructed target $\hat{x}_{1,t}$ ("reconstructed coherence") should be higher than the coherence between the target and proxy calculated for the short bivariate series. This condition is also fulfilled (Fig. 9.5b); deviations occur only at frequencies below 0.03 cpΔt, that is, for the timescales comparable to the length of the time series used for obtaining the relations between the target and proxy. Also, the true coherence in this figure and the true gain and phase factors in Fig. 9.6 were actually estimated from a time series $\mathbf{x}_t$ of length $T = 10^5 \Delta t$.

If the reconstruction results are correct, the gain and phase factors between the true and reconstructed target time series should be equal to 1 and 0, respectively, and this is what is happening (Fig. 9.6). Again, deviations from this rule occur only at the lowest frequencies where the estimates are less reliable.

Finally, the coherence function between the reconstructed time series $\hat{x}_{1,t}$ and the input $x_{2,t}$ estimated over the common interval from $t = 2$ through $t = 1350$ equals 1.00 at all frequencies different from zero (not shown). This happens because the reconstructed time series is the result of a strictly linear transformation of the input.

As the concluding step in this chapter, compare the results of time series reconstruction using the traditional correlation/regression method (CRR) and the recently

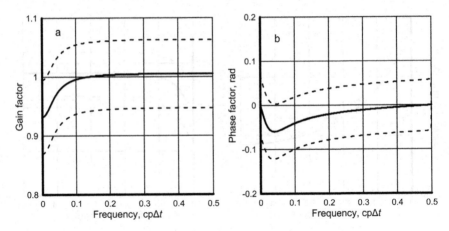

Fig. 9.6 Gain (**a**) and phase (**b**) factors of the system relating true and reconstructed time series $x_{1,t}$ and $\hat{x}_{1,t}$

published autoregressive reconstruction method (ARR), seemingly, with no reaction from time series reconstruction community in climatology, paleoclimatology, and any other Earth or solar sciences. To take into account that the maximum cross-correlation between the target and proxy time series was at the unit lag between them (see Fig. 9.3), the time series have been shifted with respect to each other. The comparison is given in Fig. 9.7 that shows the true values of the target time series (gray line) and its reconstructions through ARR (black line) and CRR (dashed line).

In complete agreement with the two approaches to reconstruction, the variances of the CRR- and ARR-reconstructed time series are 0.94 and 2.50, respectively, while the variance of the true target time series is 3.50. The CRR approach has reconstructed 27% of the target series; the latter value practically coincides with the squared cross-correlation $r_{12}(-1)$. The ARR approach has resulted in 71% of the target's series variance $\sigma_1^2 = 3.50$. The CRR-reconstructed time series is a white noise, which is starkly incorrect, while the spectral density of the ARR-reconstructed time series has the same shape as the spectrum of the true target data (see Fig. 9.5a).

Thus, the autoregressive method of time series reconstruction has been verified by using independent true data and shown to be efficient in both time and frequency domains. The case described here involves stationary random processes typical for climatology and, probably, for other Earth sciences.

9.4 Discussion and Conclusions

The traditional method of time series reconstruction based upon the cross-correlation coefficient and regression equation had been proposed at the time when theory of random processes and time series analysis were practically nonexistent. It was the

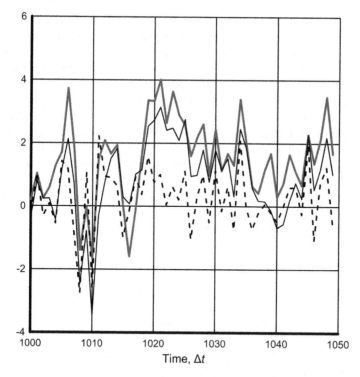

Fig. 9.7 Example of the target time series (gray) and its reconstructions with CRR (dashed line) and ARR (solid line)

only way to solve the problem at that time but it became mathematically invalid over 60 years ago, after the publication of the strategically important article by Gelfand and Yaglom (1957). A proper method of reconstruction could have been developed after the appearance of the classic book by G. Box and G. Jenkins in 1970 (the fifth edition published in 2015). The use of the CRR method with time series is hard to justify in the twenty-first century when monographs and software packages for analysis of multivariate time series written by mathematicians are easily accessible.

The autoregressive approach has been shown here to be quite powerful for detecting relations between components of bivariate time series in time and frequency domains and for time series reconstruction. The ARR method is physically sound because the autoregressive models expressed with stochastic difference equations present discrete analogs of differential equations that are used to describe and simulate geophysical processes.

The ARR method allows one to decide whether the traditional CRR approach can be applied for reconstruction of a specific time series through a single or several proxies. Strictly speaking, the regression methods works correctly only with white noise data but if the time series have identically or almost identically shaped spectra, it may still be possible to apply the traditional approach. To resolve this issue, one needs to analyze the available set of time series in the frequency domain, including estimation of scalar spectral densities, coherence functions, and coherent spectra.

The authors of an important review of temperature reconstruction methods (Christiansen and Ljungkvist 2017) justly note that "the limited knowledge of frequency-dependent characteristics in the different types of records" presents a major obstacle for understanding individual proxy records. Actually, the tools for time and frequency domain analysis of proxy and target data as scalar or multivariate time series have been available since many decades ago. They have been presented in many monographs, recommended here on numerous occasions, briefly described in Chap. 7 (and in Chap. 14 for the trivariate time series), and figured in various practical examples including analysis of proxy time series (Chap. 12). Many properties of time series are frequency-dependent and these tools can and should be used not only for time series reconstructions but also for solving many problems caused by the dependence of time series properties upon frequency.

References

Adloff F, Jordà G, Somot S, Sevault F, Arsouze C, Meyssignac B et al (2018) Improving sea level simulation in Mediterranean regional climate models. Clim Dyn 51(3):1161–1179

Ahmad S, Zhu L, Yasmeen S et al (2020) A 424-year tree-ring-based Palmer Drought Severity Index reconstruction of *Cedrus deodara* D. Don from the Hindu Kush range of Pakistan: linkages to ocean oscillations. Clim Past https://doi.org/10.5194/cp-16-783-2020

Bendat J, Piersol A (2010) Random data. Analysis and measurements procedures, 4th edn. Wiley, Hoboken

Box G, Jenkins G (1970) Time series analysis. Forecasting and control. Wiley, Hoboken

Box G, Jenkins G, Reinsel G, Liung G (2015) Time series analysis. Forecasting and control, 5th edn. Wiley, Hoboken

Bryan S, Hughen K et al (2019) Two hundred fifty years of reconstructed South Asian summer monsoon intensity and decadal-scale variability. Geophys Res Lett 46:3927–3935

Christiansen B, Ljungkvist F (2017) Challenges and perspectives for large-scale temperature reconstructions of the past two millennia. Rev Geophys 55:40–96

Douglass AE (1909) Weather cycles in the growth of big trees. Mon Wea Rev 37(6):225–237

Douglass AE (1914) A method of estimating rainfall by the growth of trees. Bull Am Geogr Soc 46(5):321–335

Douglass AE (1919) Climatic cycles and tree-growth: a study of the annual rings of trees in relation to climate and solar activity. Carnegie Inst Wash Publ 289:1

Douglass AE (1928) Climatic cycles and tree-growth. A study of the annual rings of trees in relation to climate and solar activity. Carnegie Inst Wash Publ 289:2

Douglass AE (1936) Climatic cycles and tree-growth: a study of cycles. Carnegie Inst Wash Publ 289:3

Flynn S, Knipp D et al (2018) Understanding the global variability in thermospheric nitric oxide flux using empirical orthogonal functions (EOFs). JGR Space Phys 123:4150–4170

Gelfand I, Yaglom A (1957) Calculation of the amount of information about a random function contained in another such function, Uspekhi Matematicheskikh Nauk, 12:3–52, English translation: American Mathematical Society Translation Series 2(12):199–246, 1959

IPCC (2013) Climate Change 2013: the physical science basis. Contribution of Working Group I to the fifth assessment report of the intergovernmental panel on climate change [Stocker TF, Qin

D, Plattner G-K, Tignor M, Allen SK, Boschung J, Nauels A, Xia Y, Bex V, Midgley PM (eds)]. Cambridge University Press, Cambridge, and New York, NY, 1535 pp

Lepley K, Touchan R, Melco D et al (2020) A multi-century Sierra Nevada snowpack reconstruction modeled using upper-elevation coniferous tree rings (California, USA). The Holocene, https://doi.org/10.1177/0959683620919972

Liu Y, Li C.-Y (2019) Temperature variation at the low-latitude regions of East Asia recorded by tree rings during the past six centuries. Intern J Climatol. https://doi.org/10.1002/joc.6287

Maxwell T, Silva L, Horwath W (2018) Predictable oxygen isotope exchange between plant lipids and environmental water: implications for ecosystem water balance reconstruction. J Geoph Res-Bio. https://doi.org/10.1029/2018JG00452053

Noël B, van den Berg W, Lhermitte S (2018) Six decades of glacial mass loss in the Canadian arctic archipelago. J Geophys Res-Earth. https://doi.org/10.1029/2017JF004304

Oyunmunkh B, Weijers S, Loeffler J et al (2019) Climate variations over the southern Altai mountains and dzungarian basin region, central Asia, since 1580 CE. Int J Climatol 49:4543–4558

Peng et al (2020) A 216-year tree-ring reconstruction of April-July relative humidity from Mt. Shiren, central China. Int J Climatol https://doi.org/10.1002/joc.6565

Privalsky V, Jensen D (1995) Assessment of the influence of ENSO on annual global air temperature. Dyn Atmos Oceans 22:161–178

Privalsky V (2018) A new method for reconstruction of solar irradiance. JASTP 172:138–142

Reinsel G (2003) Elements of multivariate time series analysis, 3rd edn. Springer, New York

Roy I (2018) Climate variability and sunspot activity. Springer, New York

Schweiger A, Wood K, Zhang J (2019) Arctic sea ice volume variability over 1901–2010: a model-based reconstruction. J Clim 32:4731–4752

Seftigen K, Fuentes M, Ljungqvist F, Björklund J (2020) Using Blue Intensity from drought-sensitive Pinus Sylvestris in Fennoscandia to improve reconstruction of past hydroclimate variability. Clim Dyn. https://doi.org/10.1007/s00382-020-05287-2

Shumway R, Stoffer D (2017) Time series analysis and its applications. Springer, Heidelberg

Taniguchi N, Huang C-F et al (2018) Variation of residual current in the Seto Inland Sea driven by sea level difference between the Bungo and Kii Channels. J Geophys Res-Oceans 111:C01008

Thomson R, Emery W (2014) Data analysis methods in physical oceanography, 3rd edn. Elsevier, Amsterdam

Wang J, Yang B, Osborn T et al (2019) Causes of East Asian temperature multidecadal variability since 850 CE. Geophys Res Lett. https://doi.org/10.1029/2018GL080725

Zaw Z, Zexin F, Chenxi X et al (2020) Drought reconstruction over the past two centuries in southern Myanmar using teak tree-rings: linkages to the Pacific and Indian Oceans. https://doi.org/10.1029/2020GL087627

Chapter 10
Frequency Domain Structure and Feedbacks in QBO Time Series

Abstract QBO is studied here mostly in the frequency domain as a set of six bivariate time series at atmospheric pressure levels from 10 hPa to 70 hPa, or from 31 km to 18 km above sea level. It is found to have an exceptionally stable basic frequency of 0.43 cpy with the spectra which keep their shape through the entire layer. The coherence between the time series at 10 hPa level and the other six time series down to 70 hPa is very high so that the values of coherence spectra are close to values of respective spectra especially at the peak frequency. The gain factors show the strengthening of QBO oscillations from 10 hPa to 15 hPa and 20 hPa, and then, the amplitude slowly decreases and becomes much smaller at 70 hPa. The time delay between the highest and lower levels increases from 0.2 year to 1.1 year, and the rate of downward motion decreases from 1.7 m/h to 1.3 m/h. The feedback between the levels means the existence of upward flows within the QBO system at the monthly time scales.

The goal of this chapter is to show how to analyze and understand the evolution of a bivariate time series as it progresses in space. The object of the study is the Quasi-Biennial Oscillation as it emerges at the highest level of observations at 10 hPa (31.1 km over sea level) and moves downward to the 70 hPa level at 18.4 km. The QBO as a function of altitude has been extensively studied in the time domain, but to the best of the author's knowledge, it has never been analyzed as a set of bivariate random processes in the frequency domain and this is what will be done in this chapter.

The observation data consist of seven time series given at a monthly sampling rate from 1956 through 2018 ($N = 756$). The source of data is https://www.geo.fu-berlin.de/met/ag/strat/produkte/qbo/qbo.dat (also see Chap. 5 and Naujokat 1986).

The years from 1953 through 1956 which are not available in the 10 hPa data were discarded from the other six QBO time series to allow studying the phenomenon starting from the uppermost level of observations. An illustration of QBO evolution from 10 hPa to 70 hPa levels is given in Fig. 10.1.

In accordance with Chap. 7, the following quantities that characterize the behavior of QBO in the frequency domain should be estimated for each of the six bivariate

© Springer Nature Switzerland AG 2021

V. Privalsky, *Time Series Analysis in Climatology and Related Sciences*, Progress in Geophysics, https://doi.org/10.1007/978-3-030-58055-1_10

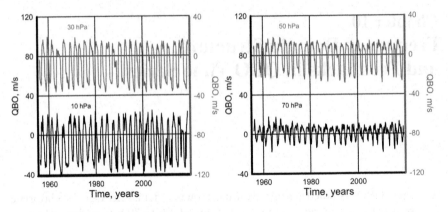

Fig. 10.1 QBO at four atmospheric levels

Table 10.1 QBO time series $x_{i1,t}$

Time series #, i	Pressure, hPa	Altitude, km	AR order p
1	10	31.1	n/a
2	15	28.4	6
3	20	26.5	4
4	30	23.8	5
5	40	22.0	7
6	50	20.6	7
7	70	18.4	6

time series beginning with QBO data at 15 hPa and ending with 70 hPa as outputs and with the 10 hPa data as the input process in each system:

- spectral densities,
- coherence function,
- coherent spectrum,
- gain factor,
- phase factor.

In addition to the above statistical functions of frequency, two more quantities are calculated for each of the six models at the QBO's frequency: the time delay corresponding to the phase factor and the rate of the QBO downward propagation.

The following notations will be used in this chapter. The time series at different atmospheric pressure levels are numbered from 1 (at 10 hPa) through 7 (at 70 hPa), as shown in Table 10.1. With the #1 time series at 10 hPa as the input, the bivariate systems will be $\mathbf{x}_{i1,t} = [x_{i,t}, x_{1,t}]', i = 2, \ldots 7$. For example, $\mathbf{x}_{51,t}$ corresponds to the linear system with $x_{5,t}$ (40 hPa) and $x_{1,t}$ (10 hPa) as the output and input processes.

Before continuing with the analysis, a few comments are needed regarding the choice of the autoregressive orders shown in Table 10.1. Normally, one should select the order recommended by the majority of order selection criteria. However, in a multidimensional case with several bivariate time series closely related to each other the task of determining the optimal order becomes more complicated. Thus, for the time series $x_{21,t}$ (15 hPa with 10 hPa), the orders $p = 2$ and $p = 6$ have been selected an equal number of times (twice), and, from statistical considerations, the lower order means relatively reliable estimates. Yet, the QBO presents a complex physical system characterized with six bivariate time series, and selecting the order for individual pairs should take into account system's complexity and the general results for the entire set. The final choice of the $\mathbf{AR}(6)$ model has been made for the time series $x_{21,t}$ having in mind the high orders selected by criteria for other bivariate systems. Choosing an $\mathbf{AR}(2)$ model would oversimplify the system.

For the time series $x_{31,t}$ (20 hPa and 10 hPa), the order $p = 4$ has been preferred by four out of the five criteria with one criterion showing the order $p = 5$. The $\mathbf{AR}(4)$ model was mandatory in this case. Besides, the differences between models $\mathbf{AR}(4)$ and $\mathbf{AR}(5)$ would not be significant.

The model $\mathbf{AR}(5)$ has been indicated for the time series $x_{41,t}$ (30 hPa and 10 hPa) by three criteria with the $\mathbf{AR}(4)$ model preferred by the other two. The choice of $\mathbf{AR}(5)$ is obvious.

The model for the time series $x_{51,t}$ (40 hPa and 10 hPa) is either $\mathbf{AR}(4)$ selected three times or the $\mathbf{AR}(7)$ selected twice. Having in mind the result for the next time series $x_{61,t}$, the final choice was $\mathbf{AR}(7)$. The choice for the time series $x_{61,t}$ (50 hPa and 10 hPa) was either $p = 7$ or $p = 2$ and the more complicated $\mathbf{AR}(7)$ model has been selected due to the disproportionate simplicity of the $\mathbf{AR}(2)$ model.

The same $\mathbf{AR}(2)$ model was preferred by two criteria for the time series $x_{71,t}$ (70 hPa and 10 hPa), and it has been ignored in order to avoid oversimplification. The other three criteria showed three different orders: 5, 6, and 11. The model $\mathbf{AR}(6)$ selected by just one criterion has been chosen as similar to the models at the higher levels. The differences between models of orders 5 and 6 are minor.

Returning to the analysis, the estimates of spectral density shown in Fig. 10.2 are obtained for the outputs of the six bivariate systems $x_{i1,t}, i = 2, \ldots, 7$ while the estimate for the time series at 10 hPa is given in the scalar version. The optimal order of the scalar model for the time series $x_{1,t}$ is $p = 11$. As mentioned in Sect. 7.3, the spectral estimates of the same time series obtained in the scalar and multivariate versions may differ and one of the reasons for those changes is the interaction between the input and output processes, first of all, of the feedbacks. This is what is happening with the spectral estimates $s_{11}(f)$ of the time series $x_{1,t}$ when it is regarded either as a scalar time series or as the input process in six different bivariate systems. The changes are not drastic, and we will not dwell on this issue here.

As seen from Fig. 10.2, the shape of the spectra is very stable and the absolute values reflect the progress of QBO with altitude: the strength of the maximum at $f \approx 0.43$ cpy does not change much with altitude staying close to 10^3 (m/s)^2/cpy until the lowest level where the QBO loses its energy by an order of magnitude.

The coherence functions $\gamma_{i1}(f)$ shown in Fig. 10.3a are extremely high even at the maximum distance of 13 km between the atmospheric pressure levels (70 hPa and 10 hPa). This high coherence at the frequencies where the spectral density achieves its

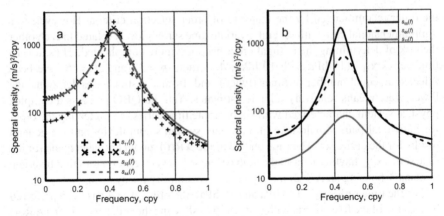

Fig. 10.2 Spectral density estimates of the output time series and the scalar spectrum $s_{11}(f)$ of the time series at 10 hPa

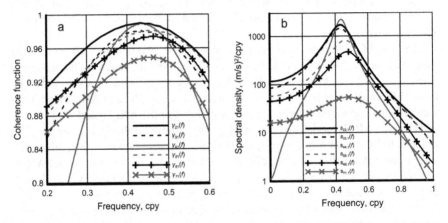

Fig. 10.3 Estimated coherence functions (**a**) and coherent spectra (**b**) of QBO autoregressive models with 10 hPa time series as the input

maximum value is rare for climatic time scales, and it shows how close the bivariate QBO time series are to strictly linear systems at least within the frequency band from 0.35 cpy to 0.50 cpy.

In accordance with the behavior of the spectra and coherence functions, the coherent spectra $s_{ii.1}(f)$, which describe the share of time series energy at the level i, which is linearly dependent upon the time series at 10 hPa having the spectrum $s_{11}(f)$, reach their maxima at frequencies close to 0.43 cpy at the upper levels and then move closer to 0.46 cpy (Fig. 10.3b). Their peak values can be as high as 98% of the spectrum $s_{ii}(f)$, $i = 2, \ldots, 7$. At all levels down to 50 hPa, the coherent spectra amount to at least 95% of respective spectra at the QBO's main frequency. The lowest percentage of 90% occurs for the system $x_{71,t}$. This behavior of the coherent spectra is the direct result of the high coherence existing within all of these bivariate systems

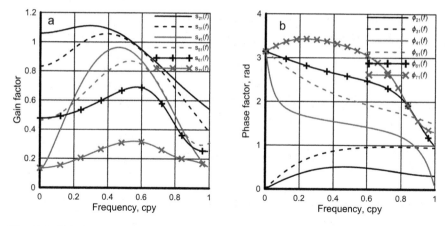

Fig. 10.4 Gain factors $g_{i1}(f)$ (**a**) and phase factors $\phi_{i1}(f)$ (**b**) for QBO autoregressive models with 10 hPa time series as the input

with $x_{1,t}$ as the input. It shows again that at frequencies close to 0.43 cpy the evolution of Quasi-Biennial Oscillations in its downward movement is very steady. It does not lose its linear character even between the highest and lowest levels of atmospheric pressure.

The frequency domain analysis produces estimates of the gain factors $g_{i1}(f)$, which show how the oscillations at 10 hPa become stronger and then weaker as they move to the lower levels. According to Fig. 10.4a, at the levels of 15 hPa and to a lesser degree, at 20 hPa the oscillations become stronger than at the top level (the gain factor $g_{i1}(f) > 1$ for $i = 2, 3$). At the lower levels, the oscillation amplitude decreases so that the response at the lowest level of 70 hPa, that is, $g_{71}(f)$, amounts to only 30% of the amplitude at 10 hPa.

The phase factor $\phi_{i1}(f)$ increases with the distance from the 10 hPa level and the value of $\phi_{71}(f)$ exceeds π at the 70 hPa level (Fig. 10.4b). This phase shift can be recalculated into the time shift $\tau_{i1}(f)$ using the relation

$$\tau_{i1}(f) = \varphi_{i1}(f)/2\pi f, \quad f > 0. \tag{10.1}$$

The changes in properties of bivariate time series at the frequency of spectral maxima for the six bivariate time series $x_{i1,t}$ are shown in Table 10.2. These frequencies differ from their analogs in the scalar cases, that is, from $f = 0.432$ (see Table 5.3), but deviations stay within 10%. The general increase of the frequency with increasing distance from the uppermost 10 hPa level shows some transformation of oscillations with altitude. Other reasons may include the cutting by three years the lengths off the six time series when switching to bivariate models and the sampling variability between scalar and bivariate estimates.

According to the table, the time required for the QBO to reach the lowest level of 70 hPa is approximately 13 months and the lag does not change much between the systems $x_{51,t}$ and $x_{61,t}$.

Table 10.2 Statistics of bivariate QBO time series with the uppermost time series as input

Time series	ΔH, km	AR order	f, cpy	$\gamma_{i1}(f_e)$	$g_{i1}(f_e)$	$\phi_{i1}(f_e)$, rad	$\tau_{i1}(f_e)$, year	$v_{i1}(f_e)$, m/h
$x_{21,t}$	2.7	6	0.431	0.99	1.08	0.50	0.19	1.66
$x_{31,t}$	4.6	4	0.428	0.98	1.05	0.89	0.33	1.58
$x_{41,t}$	7.3	5	0.436	0.99	0.96	1.52	0.56	1.50
$x_{51,t}$	9.1	7	0.442	0.99	0.84	2.15	0.78	1.35
$x_{61,t}$	10.5	7	0.460	0.98	0.66	2.60	0.90	1.33
$x_{71,t}$	12.7	6	0.473	0.95	0.29	3.33	1.12	1.29

Table 10.3 Output and input composition within QBO bivariate models, %

Time series	AR order	Output		Input	
		From itself	From input	From itself	From output
$x_{21,t}$	6	75	20	82	7
$x_{31,t}$	4	80	15	86	4
$x_{41,t}$	5	85	9	76	15
$x_{51,t}$	7	88	6	67	25
$x_{61,t}$	7	84	7	71	19
$x_{71,t}$	6	60	20	84	5

The knowledge of the time shift $\tau_{i1}(f)$ calculated from $\phi_{i1}(f)$ through Eq. (10.1) allows one to determine the rate of the downward movement of QBO. According to the results in Table 10.2, the rate is decreasing with altitude from 1.7 m/h at the top of the system to about 1.3 m/h at the bottom. Its average value is close to 1.5 m/h.

All frequency dependent characteristics in Table 10.2 present sample estimates given to the second decimal digit, and they should not be taken literally. The confidence intervals for all of them are available or can be calculated for the estimated vertical rate $v_{i1}(f)$. In particular, the conclusion about the change in the downward progress of oscillations with altitude seems to be reliable.

A surprising result has been obtained by studying the feedback between the QBO time series expressed as the relative contributions of the output processes to the inputs and vice versa. This is done using the method proposed in Chap. 7, Eq. (7.12). The QBO is a downward propagating system, and it is natural to expect some contribution of the input time series at 10 hPa to all outputs from 15 hPa through 70 hPa. This does happen though the key role is always played by the past values of the output process (Table 10.3, columns 3 and 4).

The input process, that is, the 10 hPa time series, is also formed mostly by its past values, but quite unexpectedly, it contains a noticeable contribution from three outputs: at 30, 40, and 50 hPa (Table 10.3, columns 5 and 6). An influence of a lower level upon the uppermost level at 10 hPa means a presence of feedback and, from physical considerations, it can happen only due to an upward flow within this down-ward moving system. Such a flow seems to exist in the entire system, but it becomes quite strong from the layers between 30 hpa and 50 hPa; these three processes which occur 7–10 km below the input process $x_{1,t}$ are shown to be responsible for 15–25% of its variance.

Actually, all bivariate time series belonging to the QBO system have closed feed-back loops meaning that the output process affects the input. The quantitative indica-tors of the upward flow here are the autoregressive coefficients in the input equation that reflect the contribution of the output time series to the input. As an example, consider the bivariate time series, which has the data at 70 and 30 hPa as the output and input, respectively. The optimal AR model for it is **AR(4)**:

$$x_{7,t} \approx 0.86x_{7,t-1} - 0.08x_{4,t-1} - 0.15x_{7,t-2} + 0.07x_{4,t-2}$$
$$+ 0.07x_{7,t-4} + 0.07x_{4,t-4} + a_{1,t}$$
$$x_{4,t} \approx -0.50x_{7,t-1} + 1.31x_{4,t-1} + 0.18x_{7,t-2} - 0.52x_{4,t-2}$$
$$- 0.28x_{7,t-4} + 0.08x_{4,t-4} + a_{2,t}.$$

All AR coefficients for the delay $t-3$ in this system are statistically insignificant and are not shown in the equation.

As seen from the second equation, the output process $x_{7,t}$, that is, the QBO at the lowest level of 70 hPa (12.7 km), strongly affects the input, which is 5.4 km higher than $x_{7,t}$, with three AR coefficients $\varphi_{21}^{(1)}$, $\varphi_{21}^{(2)}$, and $\varphi_{21}^{(4)}$ being statistically significant. A similar situation exists for most pairs of time series, and the first AR coefficient in the second equation is always negative. It means that the wind speed at a higher altitude is affected by the wind speed at a lower altitude and the AR coefficients $\varphi_{21}^{(1)}$ always try to change the wind direction by 180°. For $k > 1$, the AR coefficients $\varphi_{21}^{(k)}$ which reflect the influence of the lower atmospheric level upon the higher ones at time lags longer than one month may or may not be present for individual bivariate time series but the general conclusion is that in its descent from 31 km to 18 km the QBO system contains upward flows.

The origin of the flow is not clear but, according to the theoretical model by Holton and Lindzen (1972), subsequent studies of QBO (e.g., Dunkerton, 1997; Hamilton, 2010), and the review by Baldwin et al (2001), the QBO would not exist without upward vertical flows. The presence of such flows follows from the above-given analysis of QBO's structure, but the time scales of the flows found here are measured in months rather than in days as in the physical model. It would mean that the equatorial stratosphere transforms the high-frequency flows coming from the ocean to the flows whose frequency is lower by two orders of magnitude. It does not seem to be the same upward flows of planetary waves that have much shorter time scales. A more detailed analysis of the QBO phenomenon lies beyond the scope of this book.

An explicit analysis of the QBO system in the time domain is cumbersome because of the relatively high orders of bivariate models. However, the role of the Granger causality is easy to verify and it turns out that the statistical predictability of QBO time series often remains about the same in the scalar and bivariate cases, which may be explained by the relatively high AR orders chosen by order selection criteria in the scalar cases (see Sect. 5.3).

The results given here provide some seemingly new information about QBO as an almost unique case of an almost strictly linear and rather complicated geophysical system having a sharp and stable spectral peak not related to astronomical factors. The subject needs to be studied in more detail, and a physical explanation of the results is highly desired.

References

Baldwin M, Gray L, Dunkerton T et al (2001) The Quasi Biennial oscillation. Rev Geophys 39:179–229

Dunkerton T (1997) The role of gravity waves in the quasi-biennial oscillation. J Geophys Res 102(26):053–26076

Hamilton K (2010) The vertical structure of the quasi-biennial oscillation: Observations and theory. Atmos Ocean 19:236–250

Holton J, Lindzen R (1972) An updated theory for the quasi-biennial cycle of the tropical stratosphere. J Atmos Sci 29:1076–1080

Naujokat B (1986) An update of the observed Quasi-Biennial Oscillation of the stratospheric winds over the tropics. JAS 43:1873–1877

Chapter 11
Verification of General Circulation Models

Abstract Stochastic approach to verification of GCMs is advantageous because it allows one to use mathematically proper and reliable tools of analysis to test results of climate simulations, from local to global spatial scales. In this chapter, the autoregressive approach to analysis of scalar and bivariate time series is applied to test results of GCM simulations of the ENSO system, of ENSO's effect upon global temperature, and of simulated CONUS surface temperature. It is done for 47 GCM's by comparing the results of simulations with respective observation data at the annual sampling rate. Our results show that GCMs are capable of reproducing some difficult to detect features such as nonmonotonic spectra of ENSO components and degree of linear connection between them, the effect of ENSO's oceanic component on the global temperature within an intermediate frequency band with low energy, and to reproduce major statistical properties of surface temperature over the CONUS. The GCMs results are highly variable, have a tendency to overestimate the linear trend, and strongly exaggerate the effect of ENSO upon generation of annual global temperature at the most important time scales. This error compromises the GCM mechanism of generating low-frequency variations of climate.

The problem of possible anthropogenic warming of the Earth's climate has become an extremely important public issue, especially during the last several years. Suffice it to say that the total US government "expenditure for "climate science" from FY 1993 to FY 2016 come to $47.56 billion, with the international assistance amounting to $8.24 billion" (see https://www.climatedollars.org/full-study/us-govt-funding-of-climate-change/). This book is not a place to discuss the general problem of a supposedly dangerous global warming caused by human activities, but it should be remembered that warm periods lasting for decades or centuries have been observed in the past. The probability of a global warming of about 0.8 °C between 1850 and 2009 had been estimated as 0.1, which is not a small number for climate (Privalsky and Fortus 2012).

The numerical general circulation models are not capable of predicting nature-caused variations of climate at any time scales so that results of their climate simulations have to be described with a rather vague term projection. Specifically, the

© Springer Nature Switzerland AG 2021
V. Privalsky, *Time Series Analysis in Climatology and Related Sciences*,
Progress in Geophysics, https://doi.org/10.1007/978-3-030-58055-1_11

GCMs cannot provide an answer to the question of how probable it is that we are living now within just another interval of relatively warm climate caused by nature.

The efforts of scientists from many countries to evaluate climate of the current century are summarized in a fundamental report dedicated to projections of global climate including the influence of forcing functions created by anthropogenic activities (Stocker et al. 2013). Publications dedicated to verifications of those results are numerous, and this chapter presents just another contribution to the subject. Similar to other chapters, the approach here is based upon autoregressive time series analysis in time and frequency domains. Following the earlier example of using spatial averaging of simulated climate in Privalsky and Croley (1992), we will try to show that application of autoregressive time series analysis to study GCM-generated data allows one to obtain results which are easy to use for comparing observation and simulation data at a quantitative level.

A few years ago, such verifications have been conducted with 47 general circulation models developed within the framework of under the IPCC's Climate Modeling Project, Part 5 (CMIP5). The project produced records of many climate indices between 1850 and 2005, including global surface temperature and ENSO atmospheric and oceanic components SOI and NINO later analyzed in Privalsky and Yushkov 2014, 2015a, b. This chapter presents an updated version of results given in those three publications.

The CMIP5 experiments embrace practically everything related to climate variability in atmosphere, ocean, and land. In this chapter, we will discuss results of GCM verifications on three points:

- the internal structure of El Niño-Southern Oscillation system (ENSO),
- the influence of ENSO upon the global surface temperature, and
- the quality of simulations of annual surface temperature for a specific terrestrial region—the contiguous Unites States (CONUS).

This validation of climate models within CMIP5 historical experiment contains comparisons of major statistical properties of observed and simulated data at climatic time scales. The choice of the bivariate El Niño-Southern Oscillation (ENSO) system for validation purposes is based upon the fact that it presents an unusual phenomenon in the Earth's climate; therefore, the results of such validation efforts present a convenient litmus test for further validations.

The statistical properties that are estimated here to compare observed and simulated data include

- linear trend rates,
- mean values,
- variances,
- probability densities,
- autoregressive orders of stochastic difference equations that describe the time series,
- statistical predictability (persistence),
- spectral densities,

- coherence functions,
- coherent spectra.

Most observation data used in this chapter are taken from the University of East Anglia site https://crudata.uea.ac.uk/cru/data/temperature/. It means, in particular, that the results given in this chapter do not have to coincide with respective results given in Chap. 8. The time series of SOI in Chap. 8 and ASOI in this chapter as well as NINO and SST34 are similar to each other but not identical.

11.1   Verifying the Structure of ENSO

The ENSO system data have been used for validations of climate models before (e.g., Achuta Rao and Sperber 2006; Cibot et al. 2005; van Oldenborgh et al. 2005; Lin 2007), but in no case, the analyses included comparisons between all major statistical properties of observed and simulated data. Relatively recently, Guilyardi et al. (2012) analyzed 18 CMIP5 models for the ENSO region and found no "quantum leap" over the CMIP3 results for the ENSO system. Our conclusions in this section are more optimistic in some respects.

There are two well-known specific features that make the ENSO system unique. First, the spectra of both the sea surface temperature (SST) and the Southern Oscillation Index (SOI—a deseasonalized dimensionless differences of sea level pressure (SLP) between Tahiti and Darwin, Australia) behave in a manner unusual for climatology. Their energy does not decrease with growing frequency as is common for spectra of climatic variability, and their spectra are bell-shaped (e.g., von Storch 2001). The necessity for GCMs to generate time series of SST and SLP with a similar spectral density is the first definite requirement. Secondly, the two components of the system are closely related to each other: the magnitude of the linear correlation coefficient between them can be as high as 0.84 (e.g., Philander 1989) and coherence over 0.9 (Privalsky and Muzylev 2013), which is very unusual at climatic time scales. The simulated ENSO data should possess a similar property, which may be called a linear connection requirement.

The ENSO is believed to affect other processes through teleconnections over the entire globe (e.g., Alexander et al. 2002; Bartholomew and Jin 2013). Therefore, the ability of GCMs to generate time series of ENSO with the statistical properties similar to the properties of the observed data can be regarded as a necessary condition for model's acceptability. If the model fails in these two respects, the necessity of its further validation becomes doubtful. In this section, we will test the ability of GCMs used in CMIP5 to correctly reproduce major statistical properties of the actual sea level pressure and sea surface temperature data as components of the bivariate ENSO time series.

The initial observations include monthly values of SST averaged over the Niño region 3.4 (bounded by 120° W–170° W and 5° S–5° N); the data are extracted from the HadCRUT4 archive at http://www.metoffice.gov.uk/hadobs/hadcrut4/. The

SLP data at the Tahiti and Darwin stations are taken from the Australian Bureau of Meteorology site http://www.bom.gov.au/climate/current/soihtm1.shtml. The time interval for the observed and simulated data is from 1876 through 2005 (130 years).

The simulated data for the sea surface temperature (averaged over the same region) within the CMIP5 historical experiment were downloaded in accordance with the list given at the Earth System Grid Federation (ESGF) portal http://pcmdi9.llnl.gov/. The SLP data were taken from the same site. We usually analyzed only the first run of each model available for sea surface temperature and sea level pressure. The list of models is given in Table 11.1.

The validation experiment is conducted here for the annual data because we are interested in climatic time scales, which begin with year-to-year variability and do not include seasonal oscillations and other higher frequency variability. Besides, the spectral density of SST at frequencies exceeding 0.5 cpy (with the mean seasonal trend removed) is one or two orders of magnitude below the spectral density at interannual scales (see Fig. 11.1). As the mean annual values do not contain any seasonal trend, we substituted the differences of mean annual SLP values between Tahiti and Darwin for the atmospheric component of the ENSO time series and called it the annual Southern Oscillation Index (ASOI). At $\Delta t = 1$ year, ASOI is the result of a strictly linear transformation of SOI. Therefore, SST and ASOI are not the same files as NINO and SOI used elsewhere in this book.

The historical experiment was conducted within CMIP5 under the assumption that the external forcing (including the anthropogenic effects) for simulations was as

Table 11.1 List of models

Name	Name	Name
ACCESS1.0	CSIRO-Mk3.6.0	HadGEM2-ES
ACCESS1.3	CSIRO-Mk3L-1-2	INM-CM4
BCC-CSM1.1	EC-EARTH	IPSL-CM5A-LR
BCC-CSM1.1(m)	FGOALS-g2	IPSL-CM5A-MR
BNU-ESM	FIO-ESM	IPSL-CM5B-LR
CanESM2	GFDL-CM2.1	MIROC5
CCSM4	GFDL-CM3	MIROC-ESM
CESM1(BGC)	GFDL-ESM2G	MIROC-ESM-CHEM
CESM1(CAM5)	GFDL-ESM2M	MPI-ESM-LR
CESM1(FASTCHEM)	GISS-E2-H	MPI-ESM-MR
CESM1(WCCM)	GISS-E2-H-CC	MPI-ESM-P
CMCC-CESM	GISS-E2-R	MRI-CGCM3
CMCC-CM	GISS-E2-R-CC	MRI-ESM1
CMCC-CMS	HadCM3	NorESM1-M
CNRM-CM5	HadGEM2-AO	NorESM1-ME
CNRM-CM5-2	HadGEM2-CC	

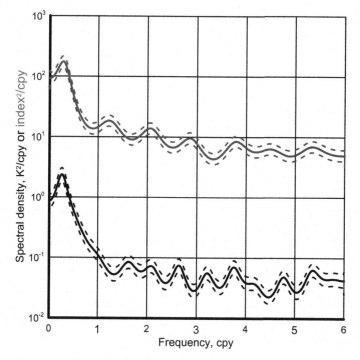

Fig. 11.1   Spectra of observed Southern Oscillation Index (gray) and sea surface temperature (black) at the sampling rate 1 month

close as possible to the observed values that had occurred between 1876 (in our case) and 2005. As the anthropogenic influence is supposed to have caused, among other things, a higher temperature of the World Ocean, it can be assumed that the trend rates differences between the observed and simulated data reflect the differences between the sensitivities of the actual climate system and GCMs to the external forcing. If the trend in the model data is significantly higher than in observations, the model is oversensitive; otherwise, it is not sensitive enough. (This is called the trend rate requirement here.)

The type of the probability density—Gaussian or non-Gaussian—was assumed in accordance with the values of the third and fourth statistical moments. If the magnitudes of standardized skewness and standardized kurtosis do not exceed 2, one can assume, at a 0.95 confidence level, that respective sample data have a Gaussian probability density (see Chap. 2 and D'Agostino and Pearson 1973). The estimation of the probability density type is important, in particular, because if the time series behave as samples of Gaussian random processes it simplifies further statistical analyses, including analysis of the time series predictability (Chap. 6). It would also mean that the results obtained within the linear Kolmogorov–Wiener theory of extrapolation cannot be improved through application of nonlinear methods (Chap. 6).

The mathematical tools used in this chapter are described earlier in this book and are based upon time and frequency domain autoregressive analysis of bivariate time series (mostly, in Chap. 7).

11.1.1   Linear Trend Rates

Sea Surface Temperature. The observed time series of SST contains a linear trend of 0.31×10^{-2} K/year, with the estimate's rms of 0.15×10^{-2} K/year. This means that the trend though statistically significant at even the 95% confidence level is close to being insignificant (and it does become insignificant if the SST time series is extended through 2012). The trend rate estimates in simulated time series vary between a statistically insignificant rate of 0.20×10^{-2} K/year (CSIRO-Mk3.6.0) and 0.93×10^{-2} K/year (IPSL-CM5A-LR) with the mean value of 0.52×10^{-2} K/year and the average rms of 0.16×10^{-2} K/year. This positive bias of more than 60% in the simulated data's trend estimate is significant. In 36 out of the 46 cases of simulated data (the file MIROC-ESM-CHEM was not available), the trend rates are faster than the trend in the observed data. On the whole, the models seem to be more sensitive to the external forcing than the actual climatic system and produce time series with positively biased linear trend.

Sea Level Pressure Difference The linear trend in the observed pressure differences (ASOI) is statistically insignificant. The simulated data reveal the same property though in several cases the trend was found to be significant (e.g., positive for EC-EARTH and negative for HadGEM2-ES). Yet, on the whole, the trend in simulated ASOI time series is not large and can be regarded as practically nonexistent. In contrast to the SST case, the models seem to behave quite properly in this respect.

11.1.2   Mean Values and Standard Deviations

Sea Surface Temperature The mean observed SST is 298.7 K with an RMS error of slightly below 0.05 K. The mean simulated SST is 299.5 K, which does not differ much from the observed mean but is still significantly different from it. The estimates of the mean SST in the simulated data vary between 297.2 K (CSIRO-Mk3.6.0) and 301.3 K (GISS-E2-R-CC) so that the range of the mean SST values as estimated by the models is 4.1 K.

The differences between the mean SST of the observed and simulated data are statistically significant in all but three cases. The average difference with the observed mean SST is approximately 0.7 K, with the maximum 2.6 K (GISS-E2-R-CC) and the minimum -1.5 K (CSIRO-Mk3.6.0). This shows a tendency to a positive bias in the simulated mean annual SST.

The standard deviation in the observed time series is 0.64 K with an RMS of 0.04 K; the average over all 46 models is 0.65 K. The two estimates are statistically equivalent. The difference between the standard deviation values of observed and simulated time series is statistically insignificant in 26 out of the 46 cases at the 95% confidence level. On the other hand, the minimum and maximum standard deviations amount to 0.4 K (MIROC-ESM) and 1.2 K (CMCC-CESM). In other words, the standard deviations of simulated data vary by 300%. This scatter in the standard deviation estimates will tell upon the variability of spectral estimates for the models and will make them less reliable.

Sea Level Pressure Difference The mean difference of the observed annual sea level pressure values between Tahiti and Darwin amounts to 2.71 mb; the respective statistics for the 46 simulated time series is 2.80 mb. The difference is statistically insignificant. The simulated mean values exhibit variability by an order of magnitude: from 0.6 mb (MRI-GCM3) to 6 mb (GISS-E2-H-CC).

The observed data have a standard deviation of 1.08 mb; the standard deviations of simulated data vary between 0.3 and 1.3 mb with the mean of 0.71 mb. The negative bias in the standard deviation estimates will cause a negative bias in the estimates of ASOI spectra.

11.1.3   Probability Density

Sea Surface Temperature The Gaussian approximation is acceptable for the probability density of the observed data and for 38 out of the 46 simulated time series. The time series whose probability distributions are significantly non-Gaussian were generated by seven models but even in those seven cases the deviations are not very spectacular. Having in mind the shortness of the data (130 annual values) and the sampling variability of estimates, it seems reasonable to assume that the simulated data can also be regarded as Gaussian.

Sea Level Pressure Difference The observed and 39 out of the 46 simulated time series have probability densities which can be regarded as Gaussian. This is an important positive property of the simulated time series of SST and ASOI.

11.1.4   Time and Frequency Domain Properties

Time Domain The best autoregressive approximation for the observed time series SST/ASOI is a model of order $p = 2$; it has been chosen by all four order selection criteria. This **AR**(2) model is given with the equation

$$x_{1,t} \approx 0.50x_{1,t-1} + 0.19x_{2,t-1} - 0.38x_{1,t-2} - \mathbf{0.08}x_{2,t-2} + a_{1,t}$$
$$x_{2,t} \approx -0.44x_{1,t-1} + \mathbf{0.06}x_{2,t-1} + 0.78x_{1,t-2} - \mathbf{0.11}x_{2,t-2} + a_{1,t} \qquad (11.1)$$

where $x_{1,t}$ and $x_{2,t}$ are the sea surface temperature SST and the atmospheric pressure difference ASOI. As usual, the statistically insignificant coefficients are shown in bold.

The innovation covariance matrix

$$\mathbf{P_a} \approx \begin{bmatrix} 0.32 & -0.47 \\ -0.47 & 0.98 \end{bmatrix}. \qquad (11.2)$$

The $\mathbf{AR}(2)$ order for the simulated data was optimal for 26 bivariate time series; 17 time series were best approximated with a Markov model $\mathbf{AR}(1)$. This covariance matrix means a high correlation coefficient between the innovation sequences: -0.85.

The bivariate time series consisting of simulated SST and ASOI may have similar or dissimilar characteristics, but comparing the time domain model given with Eqs. (11.1) and (11.2) with 46 models obtained for the simulated data would not be reasonable so we will move to the frequency domain analysis.

The observed data have two eigenfrequencies: at approximately 0.24 year^{-1} and 0.05 year^{-1}; as will be shown later, the lower frequency oscillation is not even seen in the SST and sea level pressure spectra. The eigenfrequency of about 0.2–0.3 cpy had been found for ENSO in earlier publications, both experimental (von Storch et al. 2001; Mokhov et al. 2004; Privalsky and Muzylev 2013) and theoretical (Jin 1997; Kleeman 2011). An eigenfrequency belonging to approximately the same interval is found in many time series of simulated data, though in five of the 19 cases when the optimal order of the model was $p = 1$, this oscillation was not found.

Actually, the role of the damped oscillation in the behavior of the ENSO system is quite meager because most of its energy is received through the bivariate innovation sequence a_t. As given in Chap. 6, the quantitative measure of its contribution to the variances of SST and ASOI (as well as a measure of the time series statistical predictability, or persistence) is defined through the square roots from the ratios P_{11}/σ_1^2, P_{22}/σ_2^2 where P_{11} and P_{22} are the diagonal elements of covariance matrix (11.2) and σ_1^2, σ_2^2 are the variances of SST and ASOI, respectively. For the observed temperature and atmospheric pressure differences, these ratios are close to one: $P_{11}/\sigma_1^2 = 0.82$, $P_{22}/\sigma_2^2 = 0.84$, which means that the damped periodic oscillations do not carry much energy. The annual values of SST and ASOI have low statistical predictability. Most simulated time series behave in the same manner, but about a dozen models have this predictability parameter below 0.8 with the smallest value close to 0.6. This is still not a very high statistical predictability, and it can probably be explained with the sampling variability of estimates. The share of the "deterministic" oscillation amounts to less than 20% in the observed time series and in most simulations. The average value of this parameter for the simulated SST time series is the same as for the observations. On the whole, the performance of the models seems to be satisfactory in this respect.

Some deviations of individual models from observation data, if they do not occur too often, could probably be treated as sampling variability. Essentially, we are dealing here with an ensemble of sample realizations of a bivariate random process, with each realization designed to simulate the same climatic phenomenon—natural and externally forced variability of the SST/ASOI system. Moreover, the simulated sample realizations of SST seem to be mutually independent and the same is true of the simulated SLP time series. This happens in spite of the fact that all simulated data contain the same contribution from the forcing function, which means that the predetermined forcing does not play an important role for the SST and SLP in the Niño area.

Frequency Domain Properties—Spectra As mentioned above, the scalar probability densities of SST and ASOI are usually Gaussian. This means that their bivariate probability densities are also Gaussian (Yaglom 1962) so that our autoregressive estimates of spectral matrices are the maximum entropy estimates.

According to observation data, both SST and ASOI behave rather unusually for a climate process: their spectra do not decrease with growing frequency but have a smooth and wide single maximum centered at the eigenfrequency band (see Fig. 11.2). Again, this is a previously known result obtained for scalar spectral estimates (e.g., Chu and Katz 1985; von Storch 2001; Mokhov et al. 2004) and later as bivariate maximum entropy spectra for the ENSO system (Privalsky and Muzylev 2013). As seen from Fig. 11.2, some GCMs produce time series of ASOI and SST, which are best approximated with **AR**(1) models; their behavior is different: most spectra are monotonic and some even grow with frequency, which is hardly possible for climatic processes. In those cases, the random process that generated them cannot contain any damped periodic oscillations.

The sampling variability of spectral estimates is quite large, which can hardly be regarded as a positive feature of the general circulation models.

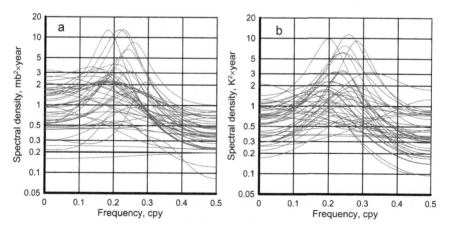

Fig. 11.2 Autoregressive (MEM) spectral estimates of simulated time series of ASOI (**a**) and SST (**b**)

The spectra of ASOI and SST estimated from observed time series and averaged over the ensembles of spectral estimates obtained from simulated data show that the average spectrum of simulated ASOI is negatively biased (Fig. 11.3a) while the SST simulations exaggerate the role of oscillations at the frequency of about 0.24 cpy (Fig. 11.3b). Having in mind the unusual structure of the ENSO bivariate time series, one may say that its simulation by the GCMs is generally satisfactory though the sampling variability of spectral estimates of simulated data is quite large.

Frequency domain properties—coherence and coherent spectra As seen from Fig. 11.4, the estimates of coherence between simulated SST and SOI show a large sampling variability and most estimates have a single smooth maximum at approximately 0.25 cpy (Fig. 11.4a). The ensemble-averaged maximum entropy estimate of

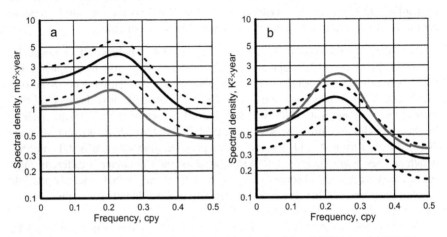

Fig. 11.3 Spectra of observed (black) and ensemble-simulated (gray) SST (**a**) and ASOI (**b**)

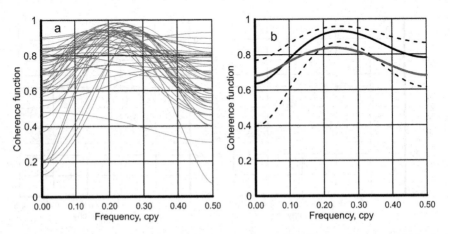

Fig. 11.4 Sample (**a**) and ensemble-averaged (**b**) autoregressive estimates of coherence between SST and ASOI. The black lines show the estimate of coherence between observed ASOI and SST

coherence between the simulated time series of ASOI and SST stays within the 90% confidence interval for the maximum autoregressive estimate of coherence everywhere but between 0.20 cpy and 0.30 cpy. The disagreement is not large and disappears if the confidence level is increased to 95%. Thus, the time series of ASOI and SST generated by GCMs are related to each other practically in the same way as the actual data of observations. This is an unquestionable success of general circulation models as a means to simulate the behavior of ENSO at climatic time scales.

Another indicator of simulation quality is the coherent spectrum. If ENSO is a system with SST and ASOI as the input and output processes, respectively, then the coherent spectrum of ASOI is the part of the total ASOI spectrum generated by the linear contribution from the input. Similarly, by regarding ASOI and SST as the input and output processes, respectively, we will have the coherent spectrum of SST as the part of the total SST spectrum that is generated by the linear contribution from ASOI.

Due to the good results with the coherence function (Fig. 11.4b) and the better results for SST with the spectra (Fig. 11.3), the results for the coherent spectra should be better for the oceanic component of ENSO, that is, for SST. This is what can be seen from Fig. 11.5. The solid black lines in the figure show the coherent spectra of ASOI and SST as outputs of bivariate time series ASOI/SST and SST/ASOI (Fig. 11.5a, b, respectively) with their 90% confidence bounds determined from observations. On the whole, the analysis of simulation results in this section shows that the general circulation models are capable, with some exceptions, of simulating main features of the El Niño—Southern Oscillation phenomenon. The exceptions include a large sampling variability within the sets of simulation data, the overestimated positive trend and negative biases in variance and spectra of simulated time series of ASOI.

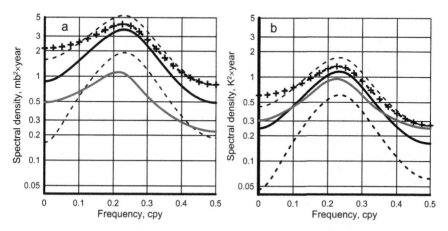

Fig. 11.5 Spectra and coherent spectra from observed time series (black lines) and from ensemble-averaged simulated estimates (gray): **a** ASOI, **b** SST

11.2 Verification of ENSO Influence upon Global Temperature

The El Niño–Southern Oscillation is believed to affect many if not all elements of the Earth climate system including the global surface temperature discussed for example, in Jones (1988) within the classical mathematical statistics and in Privalsky and Jensen (1995) within the framework of multivariate time series analysis. As high coherence between global and regional processes is a rare occurrence in the Earth's climate, such phenomena present a convenient tool for validation of climate models. In this section, the dependence of the annual global temperature upon the oceanic component of ENSO studied as a teleconnection in Chap. 8 is investigated through comparing the statistical properties of the observed system with the properties of systems simulated with the same GCMs listed in Table 11.1.

The ability of CMIP5 models to reproduce ENSO's influence upon climate characteristics has been studied in a number of publications (e.g., Langenbrunner and Neelin 2013; Tao et al. 2015; Hu et al. 2014) but the authors' approach has been through regression and correlation analysis so that their conclusions are based, to a large degree, upon thousands of estimated correlation and regression coefficients. This makes it difficult to arrive at definite conclusions about the quality of CMIP5. In this section, as in the entire chapter all comparisons are made with time series of climate elements obtained through spatial averaging. In this specific case, it is the time series of annual surface and sea surface temperature obtained, correspondingly, by averaging the observed temperature over the entire globe and over the SST34 area.

Most simulated time series of global temperature are Gaussian and estimates of their mean values proved to be quite acceptable. These are definite positive features of GCMs.

As the main goal of climate research within the framework of numerical general circulation models is predicting the anthropogenic effects upon climate, the ability of the models to correctly describe the current situation with the more or less steady growth of global temperature during the last several decades is especially important. Therefore, we will discuss the results of trend analysis by GCMs in more detail. The necessary data are given in Table 11.2.

According to these results, the mean value of linear trends estimated for all 47 models is approximately 0.48×10^{-2} K/year and its RMS is 0.18×10^{-2} K/year with the absolute values of standardized skewness and kurtosis less than 2. And though the normal probability distribution is not the best approximation for the set of linear trend estimates, there is no doubt that GCMs have generated time series of annual global temperature for the interval from 1876 through 2005 whose linear trend estimates do not disagree with the observed value of $\sim 0.53 \times 10^{-2}$ K/year and even contain a mild negative bias. This result is quite satisfactory. The trend in the time series is assumed to have been caused by an external forcing so it has been deleted from the data to reveal properties of annual global surface temperature (AGST) not influenced by the forcing.

Table 11.2 Estimated linear trend in global annual temperature according to general circulation models

Name	Trend, K/year	Name	Trend, K/year	Name	Trend, K/year
ACCESS1.0	0.2425E−02	CSIRO-Mk3.6.0	0.3041E−02	HadGEM2-ES	0.2052E−02
ACCESS1.3	0.2591E−02	CSIRO-Mk3L-1-2	0.4855E−02	INM-CM4	0.4696E−02
BCC-CSM1.1	0.6941E−02	EC-EARTH	0.6440E−02	IPSL-CM5A-LR	0.8359E−02
BCC-CSM1.1(m)	0.7013E−02	FGOALS-g2	0.4734E−02	IPSL-CM5A-MR	0.6872E−02
BNU-ESM	0.8232E−02	FIO-ESM	0.5966E−02	IPSL-CM5B-LR	0.6856E−02
CanESM2	0.5361E−02	GFDL-CM2.1	0.5735E−02	MIROC5	0.3520E−02
CCSM4	0.6896E−02	GFDL-CM3	0.2351E−02	MIROC-ESM	0.5117E−02
CESM1(BGC)	0.6698E−02	GFDL-ESM2G	0.4388E−02	MIROC-ESM_CHEM	0.4630E−02
CESM1(CAM5)	0.3670E−02	GFDL-ESM2M	0.5021E−02	MPI-ESM-LR	0.6498E−02
CESM1(FASTCHEM)	0.7368E−02	GISS-E2-H	0.5472E−02	MPI-ESM-MR	0.6693E−02
CESM1(WCCM)	0.6730E−02	GISS-E2-H-CC	0.6972E−02	MPI-ESM-P	0.6590E−02
CMCC-CESM	0.2363E−02	GISS-E2-R	0.4328E−02	MRI-CGCM3	0.3529E−02
CMCC-CM	0.3496E−02	GISS-E2-R-CC	0.4563E−02	MRI-ESM1	0.4428E−02
CMCC-CMS	0.2443E−02	HadCM3	0.3745E−02	NorESM1-M	0.4034E−02
CNRM-CM5	0.3281E−02	HadGEM2-AO	0.1857E−02	NorESM1-ME	0.3986E−02
CNRM-CM5-2	0.3800E−02	HadGEM2-CC	0.1857E−02	OBSERVED	0.5354E−02

The optimal time domain model for the observed bivariate time series AGST/SST34 is **AR**(2):

$$x_{1,t} \approx 0.55x_{1,t-1} + 0.05x_{2,t-1} + 0.26x_{1,t-2} - 0.09x_{2,t-2} + a_{1,t}$$
$$x_{2,t} \approx -0.93x_{1,t-1} + 0.30x_{2,t-1} + 0.90x_{1,t-2} - 0.33x_{2,t-2} + a_{2,t} \qquad (11.3)$$

with the covariance matrix of innovation sequence

$$\mathbf{P_a} \approx \begin{bmatrix} 0.0084 & 0.0202 \\ 0.0202 & 0.3350 \end{bmatrix}.$$

This matrix shows that the cross-correlation coefficient between the components of innovation sequence is 0.38, which means that the interdependence between the innovation sequence components plays some role in interaction between AGST and NINO.

Actually, the second equation in Eq. (11.3) differs from what had been given in Privalsky and Yushkov (2015a) in the sense that here we included the autoregressive coefficients $\varphi_{21}^{(1)}$ and $\varphi_{21}^{(2)}$ whose values lie practically on the border of respective 90% confidence intervals. The feedback between the AGST and SST34 stays weak. The feedback criteria c_{12} and c_{21} that described the improvement in linear extrapolation of AGST and SST34 at a one year lead time are equal to approximately 0.14 and 0.30, respectively, so that the criterion of feedback strength $s_{12} = c_{12}c_{21}$ amounts to just 0.04 which does not differ much from what is given for the feedback between NINO and nine sets of spatially averaged surface temperature (see Sect. 8.2).

In the frequency domain, the spectral analysis of simulated annual temperature shows fairly satisfactory results (Fig. 11.6). Many spectra lie within the 90% confidence limits for the spectrum of the observed AGST time series. Several models—CanESM2, CMCC-CESM, FIO-ESM, GFDL-CM21, GFDL-ESM2M, MIROC5, and MIROC-ESM—produce time series having statistically significant spectral maxima within the frequency band from 0.2 to 0.3 cpy; they need additional verification from their authors but can hardly change the general picture. This figure also implies that variances of simulated annual temperature more or less agree with the variance of the observed time series. A more detailed analysis shows that time series of global temperature generated by GCMs exaggerate its variance by about 20%.

Up to now, the results of analysis of simulated annual global temperature regarded as scalar time series look satisfactory in both time and frequency domains. However, the goal here was to show, if possible, the effects of the ENSO component NINO upon the annual global temperature. This can be done through analysis of the coherence function $\gamma_{12}(f)$ between AGST and NINO and through the coherent spectrum $s_{11.2}(f) = \gamma_{12}^2(f)s_{11}(f)$. The coherence functions estimates obtained from the bivariate time series of AGST and SST34 are shown in Fig. 11.7a and they can hardly be regarded as satisfactory. Obviously, there are some estimates that differ drastically from the actual coherence function (e.g., by models BCC-CSM1, CSIRO-MK3L-1-2, and GISS-E2-R-CC), but a more serious problem is the strongly exaggerated

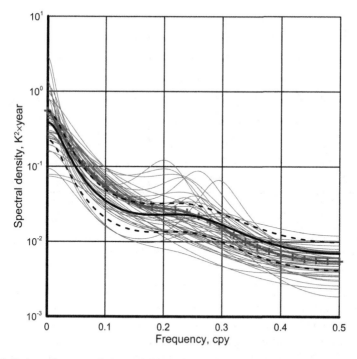

Fig. 11.6 Estimated spectra of observed (black) and simulated (gray) annual global temperature. The black line with symbols is the ensemble-averaged spectrum

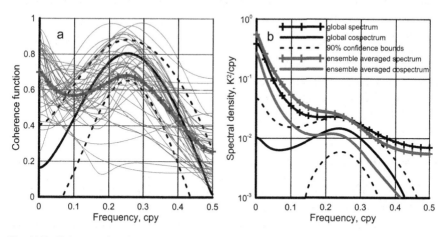

Fig. 11.7 Coherence function between AGST and SST34 (**a**) and respective coherent spectra (**b**)

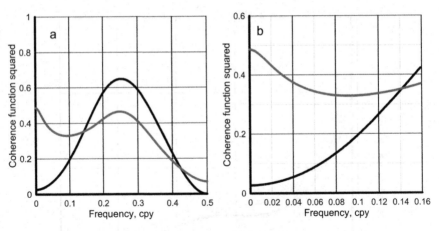

Fig. 11.8 Squared coherence function between AGST and SST34 (**a**) and its low-frequency part (**b**): results of observation analysis (black) and the ensemble-averaged estimate (gray)

coherence at frequencies below 0.14 cpy for most of the estimates obtained from simulated data. The coherence function determined from observations decreases in the low-frequency band.

These erroneous estimates of coherence result in a distorted spectrum of AGST coherent with SST34. The average over the ensemble coherent spectrum of simulated data shown in Fig. 11.7b with a gray line with symbols exceeds the estimate of coherent spectrum obtained from the observed time series of global and sea surface temperature (gray line).

The relative contribution of ENSO's oceanic component SST34 to the global surface temperature AGST as a function of frequency is measured with $\gamma_{12}^2(f)$—the coherence function squared. As seen from Fig. 11.8a, the squared average coherence obtained from analysis of simulated time series of AGST and SST34 has a negative bias within the ENSO frequency band from 0.2 to 0.3 cpy and a strong positive bias at frequencies below 0.14 cpy (at time scales longer than 7 years). The first systematic error is not very important because the spectral density of AGST in the 0.2–0.3 cpy band is small. But the large positive bias at low frequencies is disturbing because it strongly differs from what is happening in nature. Specifically, this systematic error exaggerates the ENSO's contribution by 1.5 times at $f = 0.1$ cpy and by 7 times at $f = 0.04$ cpy (Fig. 11.8b). Detailed comparisons between the observed and simulated results are shown in Table 11.3.

The degree of distortion can also be measured by comparing the contribution of NINO to the spectral density of the global surface temperature. On the whole, the contributions of the observed and simulated NINO time series to respective variances amounts to 17% for observations and 62% for simulations. At frequencies below 0.14 cpy, respective contributions are 6 and 40%. To give a specific example, the results of computations with the time series generated by CMCC-CMS model show

Table 11.3 Contribution of ENSO's oceanic component SST34 to the spectrum of global temperature in observed (obs.) and simulated (sim.) time series AGST/SST34

Frequency, cpy	Squared coherence		Frequency	Squared coherence	
	obs.	sim.		obs.	sim.
0.01	0.03	0.47	0.08	0.13	0.33
0.02	0.03	0.43	0.09	0.16	0.33
0.03	0.04	0.40	0.10	0.19	0.33
0.04	0.05	0.38	0.11	0.23	0.33
0.05	0.07	0.36	0.12	0.27	0.34
0.06	0.09	0.34	0.13	0.30	0.34
0.07	0.11	0.34	0.14	0.34	0.35

that contribution of SST34 to AGST at frequencies below 0.14 cpy extends to 61% against 7% in the observation data.

Thus, it looks like the general circulation models have some feature or features that amplify the role of NINO for the low-frequency part of the AGST spectrum. By doing this, it significantly exaggerates the effect of ENSO upon the global temperature. This behavior disagrees with observations which show a small contribution of ENSO's oceanic component to the global temperature concentrated at intermediate frequencies (also see Chap. 8). This error is especially damaging because it happens at the low-frequency end of the spectrum that describes the intensity of generation of climate variability at time scales of decades, which constitute the most important part of climate projections issued by IPCC (Stocker et al. 2013).

11.3   Verifications of Properties of Surface Temperature Over CONUS

In this section, we try to test the general circulation models for their ability to reproduce properties of annual surface temperature over a large terrestrial region—the contiguous United States, or CONUS. It includes the 48 states between Canada and Mexico and takes about 5% of the terrestrial area of the globe. The list of models analyzed here is given in Table 11.1.

Linear trend   The trend rate in the observed data amounts to 0.56 °C/100 years with the estimate's standard error 0.13 °C/100 years. The average trend rate estimate for the simulated data is 0.70 °C/100 years. More than a half of trend rate estimates for the simulated data lie outside of the 90% confidence interval for the estimate of the trend rate in the observed time series. The range of trend rate estimates in simulated data is 1.5 °C/100 years (from 0 to 1.5 °C/100 years). In other words, the trend rate in the simulated data is overestimated. Our further analysis is conducted with the time series containing no significant linear trend.

Mean values and standard deviations   The observed mean value for 1889–2005 is 11.16 °C. The average mean value estimate for the simulated data is 11.0 °C. The mean value estimates for the simulated data lie between 7.8 and 14.6 °C (a 6.8 °C range), and every one of them differs statistically significantly from the observed value. Obviously, the scatter in the mean value estimates obtained from simulated data is very large. These results are unsatisfactory.

The simulated time series has a higher variance than observations. The root mean square value (RMS) of CONUS is 0.42 °C with a 90% confidence interval between 0.38 °C and 0.45 °C. The average RMS of simulated data is 0.53 °C—a statistically significant difference from observations. On the whole, the standard deviation estimates of the CMIP5 data are positively biased.

Probability density   The observed data have a probability density function that can be regarded as Gaussian according to several criteria, including Kolmogorov–Smirnov's and chi-square. The simulated time series has the same Gaussian type of PDF with just two exceptions (IPSL-CM5B-LR and IPSL-CM5B-MR). On the whole, this is definitely a positive result.

AR model orders and persistence   The optimal model in this study was chosen for each time series with the five order selection criteria given in Chap. 3. After the linear trend removal, the observed time series over the time interval from 1889 through 2005 is best described with an AR(1) model

$$x_t = 0.21x_{t-1} + a_t. \tag{11.4}$$

with the innovation sequence variance $\sigma_a^2 = 0.17$ (°C)^2. The RMS error of the coefficient estimate in Eq. (11.1) is 0.09.

The persistence of the stochastic model (11.4) is very low: the relative predictability criterion is $\rho(1) = \sigma_a/\sigma_x \approx 0.98$. In other words, the observed time series is very close to a white noise. Most simulated time series behave in the same manner. Several simulated time series have more complicated models. For example, the time series generated with the model GFDL-ESM2M is best approximated with the AR(2) model

$$x_t \approx 0.20x_{t-1} - 0.40x_{t-2} + a_t, \tag{11.5}$$

with $\rho(1) \approx 0.92$. The predictability remains low, but the spectral density corresponding to the AR(2) model (11.5) is very different from AR(1) models.

Spectral densities   The spectral estimates obtained for the 47 simulated time series are shown in Fig. 11.9 along with the spectrum of observations (the black lines) and the average of the simulated spectra (the thick gray curve). The higher position of the average simulated spectrum happens because of the positive bias in the variance estimates. Only nine spectra of simulated data have AR orders $p = 2$ or $p = 3$

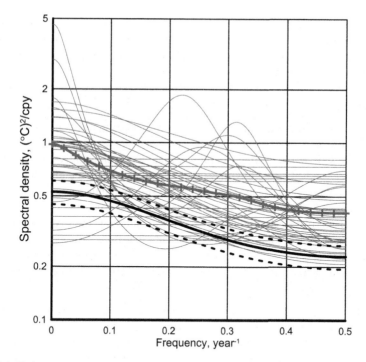

Fig. 11.9 Estimated spectra of observed (black) and simulated (gray) surface temperature over CONUS. The thick gray line with symbols is the ensemble-averaged spectrum of simulated data

and therefore can have one or two extrema or inflection points. All other spectra are constant (white noise) or monotonic (Markov chain). With few exceptions (ACESS1-0, BNU-ESM, CMCC-CMS, GFDL-ESM2M, HADGEM2-AO, MIROC-ESM), all spectra have shapes similar to the shape of the spectrum of observations. These results should be regarded as positive.

11.4 Conclusions

The general circulation models can properly describe properties of the ENSO system in some important respects but still contain a number of features that disagree with observation data. An important achievement of the models is their ability to generate time series which in most cases correctly describe the shape of the spectral density, be it the sea surface temperature, difference of atmospheric pressure between Tahiti and Darwin, annual global surface temperature or temperature averaged over the contiguous United States. Another achievement is the ability to reproduce the connections between time series in the bivariate ENSO system (ASOI and AGST). Those properties of general circulation models are important and can be regarded as relatively complicated; in this respect, the models proved their skill. Another positive feature is that most time series generated with general circulation models do not overestimate the linear trend in the global temperature.

Our analyses detected some disagreements between properties of observed and simulated data:

- the models overestimate the rate of linear trend for the oceanic component of ENSO and for the surface temperature averaged over the contiguous United States.
- the models show rather high sampling variability in practically all statistics that have been studied here; this may be a reasonable phenomenon with so many models, but it is by no means a successful result of simulations.
- the result regarding the strong influence of ENSO's oceanic component upon the spectrum of GCM-simulated global temperature at low frequencies is erroneous because according to observation data that influence occurs only within the frequency band from 0.2 cpy to 0.3 cpy where the spectral density is much smaller than at low frequencies; therefore, the total contribution of ENSO to global temperature is small.

It is the low-frequency variability of annual global temperature that plays the key role in climate behavior at all time scales longer than 7–8 years; a significant error in the composition of the spectrum of simulated global temperature is worrisome. Essentially, it means that the general circulation models that participated in CMIP5 incorrectly reproduce physical processes in the Earth climate system which are responsible for generation of climate variability. This issue requires more attention.

References

AchutaRao K, Sperber K (2006) ENSO simulation in coupled ocean-atmosphere models: are the current models better. Clim Dyn 27:1–15

Alexander M et al (2002) The atmospheric bridge: the influence of ENSO teleconnections on air–sea interaction over the global oceans. J Clim 15:2205–2231

Bartholomew H, Jin M (2013) ENSO effects on land skin temperature variations: a global study from satellite remote sensing and NCEP/NCAR reanalysis". Climate 1:53–73

Chu P, Katz R (1985) Modeling and forecasting the Southern Oscillation: a time-domain approach. Mon Weather Rev 113:1876–1888

Cibot C et al (2005) Mechanisms of tropical Pacific interannual-to-decadal variability in the ARPEGE/ORCA global coupled model. Clim Dyn 24:823–842

D'Agostino R, Pearson E (1973) Tests for departure from normality. Biometrika 60:613–622

Guilyardi E et al (2012) A first look at ENSO in CMIP5. CLIVAR Exchanges No. 58 17:29–32

Hu K et al (2014) Interdecadal variations in ENSO influences on Northwest Pacific-East Asian early summertime climate simulated in CMIP5 models. J Clim 27:5982–5998

Jin F (1997) An equatorial ocean recharge paradigm for ENSO. Part I: Conceptual model. J Atmosp Sci 54:811–829

Jones P (1988) The influence of ENSO on global temperatures. Clim Monit 17:81–89

Kleeman R (2011) Spectral analysis of multidimensional stochastic geophysical models with an application to decadal ENSO variability. J Atmos Sci 68:13–25

Langenbrunner B, Neelin J (2013) Analyzing ENSO teleconnections in CMIP models as a measure of model fidelity in simulating precipitation. J Clim 26:4431–4446

Lin J (2007) Interdecadal variability of ENSO in 21 IPCC AR4 coupled GCMs. Geophys Res Lett 34:L12702

Mokhov I, Khvorostyanov D, Eliseev A (2004) Decadal and longer term changes in El Nino-Southern Oscillation characteristics. Int J Climatol 24:401–414

Philander G (1989) El Niño and La Niña. Am Sci 77:451–459

Privalsky V, Croley T II (1992) Statistical validation of GCM-simulated climates for the U.S. Great Lakes and the C.I.S. Emba and Ural River basins. Stoch Hydrol Hydraul 6:69–80

Privalsky V, Jensen D (1995) Assessment of the influence of ENSO on annual global air temperatures. Dyn Atm Oceans 22:161–178

Privalsky V, Fortus M (2012) On possible causes of global warming. Theor Probab Appl 56:313–317

Privalsky V, Muzylev S (2013) An experimental stochastic model of the El Niño—Southern oscillation system at climatic time scales. Universal J Geosci 1:28–36

Privalsky V, Yushkov V (2014) Validation of numerical climate models for the El Niño-Southern oscillation system. Int J Ocean Clim Syst 5:1–12

Privalsky V, Yushkov V (2015a) ENSO influence upon global temperature in nature and in CMIP5 simulations. Atmosp Sci Lett 16:240–245

Privalsky V, Yushkov V (2015b) Validation of CMIP5 models for the contiguous United States. Atmosph Sci Lett 16:461–464

Stocker et al (2013) Climate change 2013: the physical science basis. contribution of working Group I to the fifth assessment report of the intergovernmental panel on climate change. Cambridge University Press, 1535 pp

Tao W et al (2015) Interdecadal modulation of ENSO teleconnections to the Indian Ocean Basin Mode and their relationship under global warming in CMIP5 models. Int J Climatol 35:391–407

van Oldenborgh G et al (2005) Did the ECMWF seasonal forecast model outperform statistical ENSO forecast models over the last 15 years. J Clim 18:3240–3249

von Storch H, Zwiers F (2001) Statistical analysis in climate research. Cambridge University Press, Cambridge

Yaglom A (1962) An introduction to the theory of stationary random functions. Prentice Hall, Englewood Cliffs

Chapter 12
Applications to Proxy Data

Abstract Normally, the proxy data are not studied in climatology as time series in time and frequency domains, and this chapter presents an attempt to begin filling up this gap. Its main goal is methodological. Eight proxy time series from ice core sites in Greenland and Antarctica are analyzed as scalar and bivariate autoregressive time series. The analysis includes estimation of PDFs, mean values, variances, correlation functions, spectral densities, coherence functions, and spectra. The time scales of analysis are from 40 years to a millennium (from 0.001 cpy to 0.025 cycles per year). Variations of ^{18}O concentration are found to have significant spatial variability; they can be approximated with low-order autoregressive models having smooth spectral densities. The Greenland time series do not show much dependence upon their past, and their spectra have a small dynamic range. In Antarctica, the dependence upon the past is strong, and the dynamic range of their spectra is large showing that processes of climate generation in Greenland and Antarctica are significantly different. The bivariate versions of time series from different ice cores show that they are not related to each other even at short distances thus showing the lack of a common climate signal.

12.1 Introduction

In climatology, proxy data are used as a basis for reconstruction of past climates. The overriding approach to climate reconstruction is the regression equation, singular (one proxy), or multivariate (more than one proxy). Irrespective of the reconstruction method, the quality of climate reconstruction depends upon the properties of proxy data. This is true for the traditional correlation/regression method based upon the mathematically incorrect approach to time series as time-invariant random vectors and for any other approach that does not lie within the framework of theory of random processes. It is also true for the autoregressive approach to time series reconstruction described in Chap. 9.

The main goal of this chapter is methodological: to show how to treat proxy data prior to transforming it into information about climate. When linear regression is used for climate reconstruction, the statistical properties of reconstructed data practically

© Springer Nature Switzerland AG 2021
V. Privalsky, *Time Series Analysis in Climatology and Related Sciences*,
Progress in Geophysics, https://doi.org/10.1007/978-3-030-58055-1_12

coincide with the properties of the proxy. This approach does not take into account that time series have to be studied in both time and frequency domains and that statistics of properly reconstructed time series do not necessarily reproduce statistics of the proxy. Knowing properties of the proxy in time and frequency domains is mandatory, and it will be shown here what information can be extracted from the proxy data for climate reconstruction using, as examples, the data on variations of oxygen isotope ^{18}O concentration obtained from ice cores in Greenland and Antarctica.

Effective methods of time series analysis have been developed decades ago and are described in a number of books popular in science and engineering but barely known and rarely used in climatology and related sciences. The important sources of information required for studying paleoclimatic time series include the books authored by Granger and Hatanaka (1964), Bendat and Piersol (2010), Box et al (2015), and Shumway and Stoffer (2017).

The spectra of ice core data have been examined previously (e.g., Masson et al. 2000; Watanabe et al. 2003) but a study of ice core data as scalar and bivariate time series taken in this chapter does not seem to have been undertaken before.

If the goal of reconstruction is to restore past climate as a function of time, for example, to obtain a time-dependent sequence of temperature using a sequence of ice core data or annual tree ring widths as a proxy, we are dealing with time series— time-dependent sequences of random variables, or random functions of time. Consequently, in order to reconstruct the behavior of some climate variable into the past, one needs to have a proxy or proxies in the form of time series. The properties of the proxy should be described with statistics such as probability density function, mean value, variance, covariance function, and spectral density. When studying relations between proxies or between a proxy and a climate index for subsequent reconstructions, the required information becomes more complicated and should include bivariate (multivariate, if there are more than one proxy) probability densities, vectors of mean values, covariance and spectral matrices, and some functions of frequency such as coherence functions and coherence spectra.

If the proxy time series cover long time intervals, for example, several millennia, one needs to know whether the time series can be regarded as samples of stationary random processes over the entire time interval of proxy observations or over its subintervals. Using a single equation for reconstructing temperature over the entire interval may not be correct because relations between the proxy and temperature may have been different for different climate epochs. This seems to be especially relevant in paleoclimatology where the epochs of more or less steady behavior, slow growth or decrease of the proxy may be just parts of the same time series. The knowledge of time series properties is necessary for any type of proxy data and for any method of reconstruction, be it the traditional linear regression, the autoregressive reconstruction described in Chap. 9, or any other approach.

The concept of time series had not been known to the founder of dendrochronology Andrew Douglass, which is quite understandable for the first part of the previous century, but the term time series is widely used in modern paleoclimatology (e.g., Bradley 2015 or White et al. 2018). Whenever the term is applied to research data, it automatically means that methods of analysis must be based upon the theory of random processes rather than upon the classical mathematical statistics, which has nothing to do with time series.

The proper tools of analysis have already been described in the previous chapters so that it will be sufficient now to list major stages of analysis and give some examples. Those stages are listed below under the assumption that the time series of proxies used for subsequent research meet the requirements to data processing and do not need any preliminary analysis.

1. Scalar analysis: each of the time series of proxy variables is analyzed in accordance with the methods described in the first six chapters of the book: determine whether the time series can be regarded as a sample of a stationary random process and conduct time and frequency domain analysis to assess and examine its basic statistics: probability density function, mean value and variance with proper confidence intervals, an optimal autoregressive model, and respective spectral density.

2. Bivariate analysis: finding an optimal time domain model and calculating frequency domain quantities—spectra and coherent spectra, coherence functions, and, if necessary, the noise spectrum and gain and phase factors. The analysis will show what information the time series carry about each other and will provide estimates of respective frequency-dependent functions. Examples can be found in Chaps. 8, 9, and 10. Clearly, in the case of ice core data, any interdependences between time series are generated by a common forcing function—variations of external conditions including climate.

This chapter presents an attempt to introduce methods of scalar and bivariate time series analysis in time and frequency domains into paleoclimatology providing tools that may help to answer some unanswered questions involving proxy data and climate. The major task discussed in this chapter is how to reveal the probabilistic properties of proxy time series (ice core data in this case) that are used to reconstruct temperature or some other climate variable at time scales from decades or centuries to millennia.

The intention here is to show what specific information can be extracted from the results of probabilistic analysis conducted in agreement with theory of random processes and hopefully to convince the researcher that the approach is helpful and even mandatory for getting and understanding the information contained in time series of proxy data. The sampling interval in the Greenland and Antarctic data is 20 years so it is natural to expect that the time series covering millennia can give us information about climate variability at time scales from, roughly, 50 years to a millennium. These are the expected approximate limits of time scales in the analysis of data in this chapter.

12.2   Greenland Ice Cores

The Web site https://www1.ncdc.noaa.gov/pub/data/paleo/icecore/greenland/vinthe r2009greenland.txt contains detailed information about ice core locations in Greenland and Northern Canada; a full description of the entire project is given in publications by Vinther et al (2008, 2009). The sites used in this section include A84/87 (Agassiz), Renland, GRIP, and NGRIP. Respective basic data are given in Table 12.1. The time series consist of variations of oxygen isotope anomalies $\delta^{18}O$ measured in per mille (‰). They are synchronized and extend from 40 years to 11,700 years prior to year 2000 at the sampling interval $\Delta t = 20$ years. The full length of each time series is $N = 584$ or $T = N\Delta t = 11680$ years.

The Agassiz/Renland pair is selected here for further analysis because the time series have been used by Vinther et al. (2009) for temperature reconstruction; the distance between those two sites is close to 1600 km. The distance between GRIP and NGRIP barely exceeds 300 km; the short distance should mean a close relation between the time series, and we intend to study it.

As seen from Fig. 12.1, all four time series contain a sharp decrease of isotope concentration at about the 10 ka (millennia) point, which makes it impossible to study the entire time series as belonging to a single sample of some stationary random process. Therefore, we will begin with analyzing the first 10 ka of the data which look like samples of a stationary process plus a linear trend.

Table 12.1 Coordinates and altitudes of ice core sites in Greenland

#	Name	Notation	Coordinates	Meters above mean sea level
1	Agassiz	$x_{1,t}$	80.7° N, 73.1° W	1730
2	Renland	$x_{2,t}$	71.27° N, 26.73° W	2350
3	GRIP	$x_{3,t}$	72.58° N, 37.64° W	3230
4	NGRIP	$x_{4,t}$	75.10° N, 42.32° W	2917

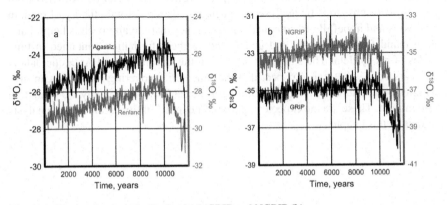

Fig. 12.1 Time series Agassiz, Renland (**a**), GRIP, and NGRIP (**b**)

The length of the time series from the beginning to the point of the sharp decrease is $N = 499$ sample intervals, that is, $T = N\Delta t = 9980$ years starting with year 40. The probability densities of the time series are definitely not normal so the advantage of a Gaussian distribution does not exist for the set of isotope concentration as a whole. It does not seem to be a great disadvantage because the task of forecasting is not relevant in this case. On the other hand, being non-Gaussian, they have to be considered as nonergodic.

Our task here is to understand basic statistical properties of isotope time series including their behavior in time and frequency domains as described with autoregressive stochastic models and respective autoregressive (or maximum entropy) spectral density estimates. The presence of a trend in the time series may mask their features at time scales of a millennium and shorter so that our results will be given for the time series that do not contain any linear trend (Fig. 12.2).

The information about the possible statistical equivalence of mean values and variances for the time series is obtained through the autoregressive analysis of respective scalar time series; its results include confidence intervals for the first two statistical moments of each time series. As seen from Table 12.2, the confidence intervals for estimates of mean values for each member of the pair Agassiz/Renland do not overlap at a confidence level 0.90; however, they do at the 0.95 confidence level (not shown). The confidence intervals for variance estimates overlap so that the time series Agassiz and Renland may belong to the same stationary random process; this

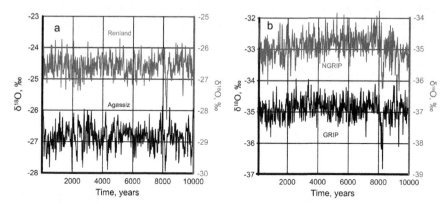

Fig. 12.2 Time series without the linear trend

Table 12.2 Estimated mean values and variances with 90% confidence limits	Time series	Mean values, ‰	Variances, (‰)^2
	Agassiz	−26.89 −26.85 −26.81	0.122 0.138 0.157
	Renland	−26.61 −26.58 −26.55	0.097 0.109 0.122
	GRIP	−34.94 −34.90 −34.86	0.115 0.129 0.145
	NGRIP	−34.98 −34.93 −34.88	0.141 0.160 0.181

statement is also true for the time series GRIP and NGRIP. This information is useful for evaluating the statistical equivalence of time series in each pair.

With the time series that cover intervals of many centuries, such test should be conducted cautiously having in mind possible low-frequency variations of climate. If the results of the tests do not show serious discrepancies in statistical properties, one should ease the requirements, at least at the initial stage of the study. If the changes are really significant, the analysis will have to be done separately for individual parts of the time series with the goal to understand what was happening with it.

The optimal time domain models for the Agassiz and Renland scalar time series are Markov, that is, AR(1) processes:

$$x_{1,t} \approx 0.43 x_{1,t-1} + a_{1,t}$$

and

$$x_{2,t} \approx 0.31 x_{2,t-1} + a_{2,t}.$$

Here, $x_{1,t}$ and $x_{2,t}$ are the Agassiz and Renland time series and $a_{1,t}, a_{2,t}$ are respective innovation sequences.

The estimated AR coefficients are significantly different from zero, and the confidence intervals for them overlap meaning that the difference between the two models is not crucial. In a Markov model, the value of the AR coefficient φ defines the persistence criterion, that is, $r_e(1) = \varphi$ (0.43 and 0.31 in our case). The persistence criterion characterizes the degree of time series dependence of upon its past and for the time series $x_{1,t}$ and $x_{2,t}$ the dependence is rather low (see Chaps. 3 and 6).

The best models for the neighboring time series GRIP and NGRIP are autoregressive sequences of order $p = 3$:

$$x_{3,t} \approx 0.14 x_{3,t-1} + 0.17 x_{3,t-2} + 0.12 x_{3,t-3} + a_{3,t}$$

and

$$x_{4,t} \approx 0.22 x_{4,t-1} + 0.15 x_{4,t-2} + 0.14 x_{4,t-3} + a_{4,t}.$$

These time series remember their past for three sampling intervals, that is, for 60 years, and the contribution of the past to the current value is distributed more or less evenly from year to year. The variances of innovation sequences $a_{3,t}$ and $a_{4,t}$ are 0.118 (‰)^2 and 0.139 (‰)^2, respectively; with the values of variances $\sigma_3^2 \approx 0.129$ (‰)^2 and $\sigma_4^2 \approx 0.160$ (‰)^2, the persistence criteria $r_e(1) = \sqrt{1 - \sigma_a^2/\sigma_x^2}$ stay below 0.4, that is, these time series also have a weak memory.

The spectral density estimates obtained through the four above given autoregressive models are shown in Fig. 12.3. In agreement with the time domain models, the Agassiz and Renland spectra diminish monotonically but the GRIP and NGRIP spectra show statistically significant maxima at frequencies between 0.017 and 0.019

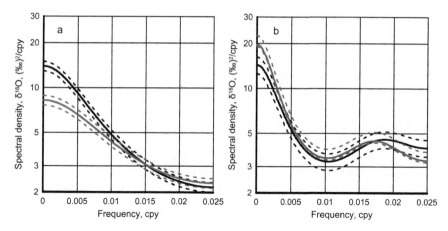

Fig. 12.3 Autoregressive (maximum entropy) estimates of time series: **a** Agassiz (black) and Renland (gray), **b** GRIP (black) and NGRIP (gray)

cpy corresponding to the time scales of about 50–60 years. These maxima may play a noticeable role in the time series behavior at high frequencies.

If temperature is reconstructed from the time series of isotope concentrations through a linear regression equation, the behavior of the Agassiz and Renland time series of $\delta^{18}O$ agrees with the assumption that a Markov model with a small AR coefficient can serve as a good approximation for climate variability (Hasselmann 1976). In this case, the hypothesis is confirmed at time scales from decades to millennia (also see Dobrovolski 2000). The Markov model is not optimal for all four time series but the approximation works for the GRIP and NGRIP time series as well.

The availability of several time series of isotope concentration makes it possible to examine relations between them. A time series at any site may be found related to a time series at another site due to a common external forcing which may include climate.

The forcing can be detected if the time series are long enough for studying their interdependence and in our case the available data set meets this requirement. Studying the bivariate data sets, that is, determining dependence between the time series in time and frequency domains, will begin here with the two ice core sites which are closest to each other, that is, with GRIP and NGRIP.

The optimal bivariate autoregressive model for this pair is **AR**(2):

$$x_{3,t} \approx 0.15x_{3,t-1} + \mathbf{0.01}x_{4,t-1} + 0.14x_{3,t-2} + 0.10x_{4,t-2} + a_{3,t}$$
$$x_{4,t} \approx \mathbf{0.03}x_{3,t-1} + 0.23x_{4,t-1} + 0.11x_{3,t-2} + 0.14x_{4,t-2} + a_{4,t}. \qquad (12.1)$$

Here, as always in this book, the numbers in bold do not differ significantly from zero.

The covariance matrix of the bivariate white noise $\mathbf{a}_t = [a_{3,t}, a_{4,t}]'$ is

$$\mathbf{P_a} \approx \begin{bmatrix} 0.12 & 0.05 \\ 0.05 & 0.14 \end{bmatrix}, \tag{12.2}$$

which means that the cross-correlation coefficient between the components of $\mathbf{a}_t$ is 0.41. The model described with Eqs. (12.1) and (12.2) has been indicated by three of the five order selection criteria used here.

The presence of statistically significant AR coefficients $\varphi_{34}^{(2)} \approx 0.10$ and $\varphi_{43}^{(2)} \approx$ 0.14 as well as of statistically insignificant coefficients $\varphi_{34}^{(1)} = 0.01$ and $\varphi_{43}^{(1)} = 0.03$ in the first and second equations of Eq. (12.1) shows that both time series are influenced by a common force and that it takes two sampling intervals, that is, 40 years for the effect of the external force upon one time series to be seen in the other one.

The bivariate time domain model given with Eq. (12.1) does not carry any explicit quantitative information about the forcing function that is common for both GRIP and NGRIP time series but such information can be obtained in the frequency domain. In their recent work, Münch and Laepple (2018) suggested a method of climate signal detection that, according to the authors, worked well on interannual and decadal time scales (that is, their low frequencies are the highest frequencies in our case) but was less efficient at lower frequencies. The authors have reasonably assumed that each time series carries a climate signal and a noise caused by other factors, with both signal and noise being frequency dependent. In their attempt to separate the common climate signal from noise, they suggested a method based upon stacking time series that carry information about the effect of climate upon isotope history in ice core data obtained at different sites. Actually, the tools for solving this task had been known in information theory and multivariate time series analysis since decades ago (Gelfand and Yaglom 1957; Granger and Hatanaka 1964; Bendat and Piersol 2010). As shown in Chap. 7, the required functions of frequency in the bivariate case are the coherence function $\gamma_{ij}(f)$ and the coherent spectrum $s_{ii.j}(f) = \gamma_{ij}^2(f)s_{ii}(f)$. The noise spectrum can be calculated as $n_{ii.j}(f) = s_{ii}(f) - s_{ii.j}(f)$ and it is easy to show that the signal-to-noise ratio is found as $\gamma_{ij}^2(f)/[1 - \gamma_{ij}^2(f)]$. Respective quantities in the multivariate case are the multiple coherence and multiple coherent spectra (see Chap. 14).

Consider now an example of detecting a climate signal in the time series GRIP regarded as a component of a bivariate time series GRIP/NGRIP, that is, $\mathbf{x}_t = [x_{3,t}, x_{4,t}]'$. As it has already been described just above, the best bivariate time domain model for the time series GRIP/NGRIP is an $\mathbf{AR}(2)$. Its frequency domain version includes the spectral densities $s_{33}(f)$, $s_{44}(f)$, the coherence function $\gamma_{34}(f)$, and the coherent spectrum $s_{33.4}(f)$. The noise spectrum $n_{33.4}(f)$ is $n_{33.4}(f) = s_{33}(f) - s_{33.4}(f)$.

The spectral density estimates of GRIP and NGRIP time series obtained through the bivariate time domain model given with Eqs. (12.1) and (12.2) are shown in Fig. 12.4. The spectral maximum at about 0.017 cpy, seen previously in the GRIP and NGRIP spectra, has disappeared as a result of going from a scalar model to a bivariate one. (This feature of multivariate time series has been noted in Chap. 7).

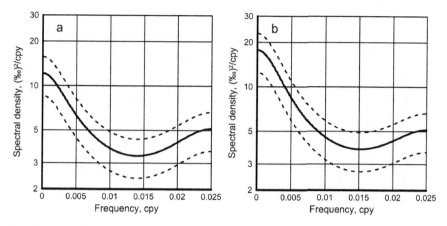

Fig. 12.4 Spectra of GRIP (**a**) and NGRIP (**b**) time series according to the bivariate model (12.1)

The short (~300 km) distance between the ice core sites allows one to presume that they would be closely correlated with each other because climate variations at two points so close to each other and at about the same elevation could hardly be very different. However, the cross-correlation function between $x_{3,t}$ (GRID) and $x_{4,t}$ (NGRID) does not show a close relation (Fig. 12.5). As the cross-correlation function is by no means a convenient tool for detecting dependence between time series, the analysis should be continued to the coherence function which describes relations between time series as a function of frequency.

Fig. 12.5 Cross-correlation function between GRIP and NGRIP time series

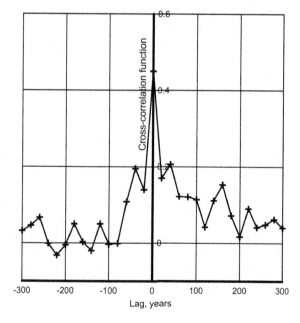

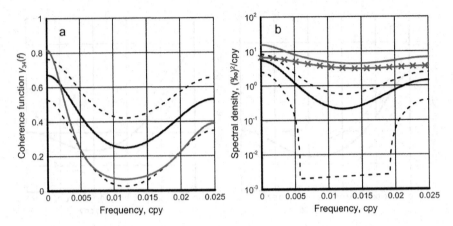

Fig. 12.6 Signal-to-noise ratio (gray) and coherence function $\gamma_{34}(f)$ **(a)**, coherent spectrum $s_{33.4}(f)$ and noise spectrum $n_{33.4}(f)$, (black line and gray line with symbols) **(b)**. The gray line shows the GRIP spectrum $s_{33}(f)$

As seen from Fig. 12.6a, the coherence is statistically significant over the entire frequency band and is higher at high and low frequencies. But it is quite low at frequencies between 0.005 and 0.02 cpy, that is, at time scales from 50 to 200 years. In other words, the time series do not seem to carry much information about each other at the frequency band most important for studying climate variability at intermediate time scales.

The signal-to-noise ratio (the gray line in Fig. 12.6a) is small everywhere with the exception of the lowest frequencies where the time scales are comparable to the time series length; the results that can be obtained for that frequency band are relatively unreliable. Both the signal-to-noise ratio and coherence function curves repeat the shape of the GRIP spectral density $s_{33}(f)$. The estimated coherence spectrum $s_{33.4}(f)$ in Fig. 12.6b (black lines) is significantly smaller than the GRIP spectrum $s_{33}(f)$ shown with a gray line. The strength of the common signal as a function of frequency is characterized with the ratio $s_{33.4}(f)/s_{33}(f) = \gamma_{12}^2(f)$. It amounts to 20–45% at frequencies below 0.005 cpy (time scales 200 years and longer), 18–28% at frequencies above 0.02 cpy (not longer than 50 years), and drops to 6–18% between 0.010 and 0.015 cpy (time sales from 70 years to 100 years). In other words, the common signal is weak at all time scales significantly shorter than the time series length and very weak within the interval from decades to several centuries. As the result, the noise spectrum $n_{33.4}(f)$ (gray line with symbols) is close to the GRIP spectrum $s_{33}(f)$. The coherence function describes information exchange for both time series so that the above conclusion holds for the NGRIP spectrum $s_{44}(f)$.

These results actually mean that the climate signal is strong only at frequencies which corresponds to the time scales over 400–500 years (the gray line in Fig. 12.6a). Moreover, as seen from Fig. 12.6b, the noise spectrum $n_{33.4}(f)$ (the gray line with symbols), which is the part of the spectrum $s_{33}(f)$ not related at all to the climate signal, exceeds the coherent spectrum—the contribution of the climate signal—within the entire frequency band. It looks like the time series of oxygen isotope ^{18}O variations if it is to be transformed into temperature through a linear regression equation would hardly bring much information about climate variability at time scales from decades to about half a millennium. Thus, the results of analysis of the bivariate time series GRIP/NGRIP show that in spite of the short distance between the sites the common signal in them is weak and becomes visible only at very low frequencies while the bulk of the time series spectra consists of contributions from noise. Note that results of analysis obtained for time scale comparable to the time series length cannot be reliable.

The components of the bivariate time series Agassiz/Renland behave in agreement with the Hasselmann's hypothesis of climate variations: the best scalar model for each is a Markov model AR(1) with autoregressive coefficients of about 0.2–0.4. The optimal bivariate model is **AR**(1)—a bivariate Markov process. For the Agassiz/Renland time series, it can be written as

$$x_{1,t} \approx 0.40 x_{1,t-1} + a_{1,t}$$
$$x_{2,t} \approx 0.23 x_{2,t-1} + a_{2,t} \tag{12.3}$$

The terms that contain off-diagonal coefficients $\varphi_{12}^{(1)}$ and $\varphi_{21}^{(1)}$ that should be present, respectively, in the first and second equation do not appear because they are very close to zero. Moreover, the covariance matrix of the bivariate noise $\mathbf{a}_t = [a_{1,t}, a_{2,t}]'$ is almost diagonal which means the lack of correlation between the innovation sequences. Thus, the long-term variations of climate at these two sites divided by a distance of about 1600 km and at altitudes 2000 ± 300 m in the polar latitudinal zone from $70°$ N to $80°$ N behave as samples of two mutually uncorrelated Markov processes. It means, in particular, that the temperature variations at the Renland ice core area at $71°$ N and Agassiz site at $81°$ N can be completely asynchronous even at time scales of millennia. It also means that the time series do not contain any common signal.

The end parts of the time series after the linear trend removal from them are shown in Fig. 12.7. Their analysis as scalar time series leads to more or less the same results as in the case of the longer parts: the best models are Markov processes with the autoregressive coefficient of 0.3–0.6. The behavior of the bivariate models is different. The statistical dependence within the pair Agassiz/Renland does not exist but the neighboring sites GRIP and NGRIP show some interdependence caused, of course, by the external forcing. The **AR**(1) model for this time series is

$$x_{1,t} \approx 0.31 x_{2,t-1} + a_{1,t}$$
$$x_{2,t} \approx 0.40 x_{2,t-1} + a_{2,t} \tag{12.4}$$

with the innovation covariance matrix

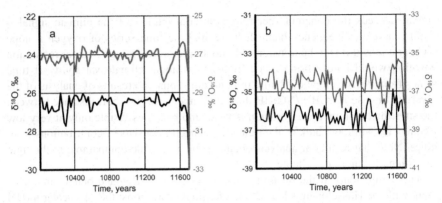

Fig. 12.7 End parts of the time series: Agassiz (black) and Renland (gray), GRIP (black), and NGRIP (gray)

$$\mathbf{P_a} \approx \begin{bmatrix} 0.19 & 0.12 \\ 0.12 & 0.26 \end{bmatrix}, \tag{12.5}$$

which correspond to a rather high correlation coefficient of 0.55 between the white noise components. The influence of the past value $x_{2,t-1}$ upon $x_{1,t}$ and the correlation between the white noise components result in a high values of coherence function $\gamma_{12}(f)$, which exceeds 0.8 at lower frequencies and decreases monotonically to 0.016 cpy becoming statistically insignificant at higher frequencies. Generally, it looks like the results for the shorter bivariate time series Agassiz/Renland and GRIP/NGRIP do not qualitatively contradict previous results for the longer time series but with these short time series obtaining reliable results for the bivariate case is doubtful.

In concluding this section, one may say that results of analysis obtained from the longer bivariate time series from 40 years to 10,000 years prior to year 2000 and at least to some extent with the shorter time series from 10,020 years to 11,700 years demonstrate the lack of a substantial climate effect upon variations of ^{18}O isotope anywhere but at the lowest frequencies.

12.3   Antarctic Ice Cores

The initial data in this case consist of five time series of oxygen isotope δ^{18}O concentration in five ice cores taken in Antarctica. Each time series carries information about the δ^{18}O concentration between 9 and 22 ka prior to 1950 at a sampling interval $\Delta t = 20$ years. The data are available at the site ftp://ncdc.noaa.gov/pub/data/paleo/icecore/antarctica/antarctica2011iso.txt. In this section, we will concentrate upon time and frequency domain analysis of the data presented with four of these time series. The paleoclimatology issues related to the ice cores are given in the article by Pedro et al. (2011) and cannot be discussed here.

The names, geographical coordinates, and altitudes of the sites over the seal level are shown in Table 12.3. Thus, the ice core locations lie within the area limited by approximately 70° S–80° S and 100° E–160° W, so that distances between the locations vary from 600 km to 3500 km. The altitude differences amount up to approximately 2000 m. We assume that the time series given at the above Web sites and described in Pedro et al. (2011) can be regarded as potential indicators of climate variations in Antarctica.

As seen from Fig. 12.8, the trajectories of the $\delta^{18}O$ time series are rather complicated. Our goal is to see if statistical properties of the time series which have been produced within this relatively small continent behave in a similar manner and, if they do, would the information about climate variations obtained as the result of analysis be similar for these time series. Having acquired some experience with the Greenland ice core data, we will generally follow the same approach to analyzing the same type of data obtained in Antarctica.

Table 12.3 Coordinates and altitudes of ice core sites in Antarctica

#	Name	Notation	Coordinates	Meters above mean sea level
1	Byrd	$x_{1,t}$	80°01′ S 119°31′ W	1530
2	EDML	$x_{2,t}$	75°00′ S 00°04′ E	2892
3	Law Dome	$x_{3,t}$	66°46′ S 112°48′ E	1370
4	Siple Dome	$x_{4,t}$	81°40′ S 148°49′ W	621
5	Talos Dome	$x_{5,t}$	72°49′ S 159°11′ E	2315

Fig. 12.8 Time series of $\delta^{18}O$

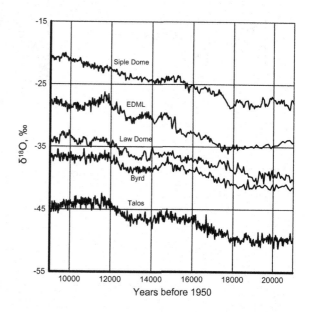

Following the example with the Greenland data, we will study four time series: at Byrd ($x_{1,t}$), Siple Dome ($x_{4,t}$), EDML ($x_{2,t}$), and at Law Dome ($x_{3,t}$). The pairs Byrd/Siple Dome and EDML/Law Dome are selected because the distance between the components of the pairs is the shortest (~550 km) in the first case and the longest (~3500 km) in the second case. Besides, the part of the Siple Dome time series after 18 ka looks very different from all other data so that the end point of each time series was set to 18 ka. This means that the length of the time series $N = 450$ or $T = 9000$ years. The selected time series with no linear trend are shown in Fig. 12.9. The Gaussian approximation for their PDFs does not work.

As in the Greenland case, the confidence intervals for estimates of mean values and root mean square values given in Table 12.4 are calculated using the numbers of independent observations determined through the correlation functions built on the basis of time domain autoregressive models fitted to the time series (see Chap. 4). They do not overlap for the mean value estimates and overlap for the variance estimates for the pair EDML/Law Dome. We will have to assume that the lack of overlap is caused by different climates at the ice core sites even when the distance between them is relatively short. The difference in elevations (roughly 1000–1500 m) must also have some influence.

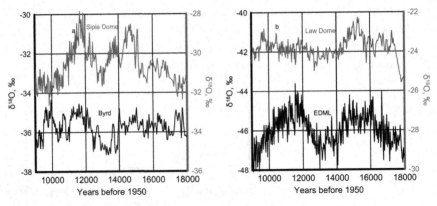

Fig. 12.9 Time series of $\delta^{18}O$ without the linear trend

Table 12.4 Estimated mean values and variances with 90% confidence limits	Time series	Notation	Mean values (‰)	Variances (‰)^2
	Byrd	$x_{1,t}$	−37.85 −37.63 − 37.41	0.469 0.570 0.691
	EDML	$x_{2,t}$	−46.46 −45.99 − 45.52	0.677 0.821 1.049
	Law Dome	$x_{3,t}$	−26.26 −25.93 − 25.60	0.553 0.677 0.879
	Siple Dome	$x_{4,t}$	−31.10 −30.58 − 30.06	0.732 0.914 1.228

Table 12.5 Parameters of
time domain models

#	Name	AR order	AR coefficients
1	Byrd	3	1.36 −0.70 0.28
2	EDML	7	0.35 0.05 0.18 ...
3	Law Dome	3	0.95 −0.19 0.16
4	Siple Dome	3	0.74 0.03 0.17

In contrast to what has been obtained for the Greenland data and to the assumption about the behavior of climate as a Markov process with a small autoregressive coefficient, the models of Antarctica time series have high autoregressive coefficients (Table 12.5) and what is more important, their spectra quickly decrease with frequency.

The autoregressive spectral estimates obtained from the time series (Fig. 12.10) reveal random processes whose properties differ very much from what has been learned from the Greenland ice cores. (The confidence intervals for the estimates are not shown because the spectra are practically monotonic.) Generally, the "noisy" part of the spectra begins at 0.005 cpy, that is, at time scales shorter than 200 years rather than at 100 years and less as noticed in Pedro et al. (2011). The time interval of 200 years contains ten observed values of isotope concentration at the sampling interval $\Delta t = 20$ years, and the total length of 12 ka contains 600 such time intervals which is more than enough for getting reliable estimates of spectral density at higher frequencies. If we believe that variations of the time series with frequencies higher than 0.005 cpy are generated by noise, the "signal" part of the spectrum consists only of time scales over 200 years, which is hardly good enough for a long time series with a sampling interval $\Delta t = 20$ years. Moreover, at frequencies higher than 0.005 cpy, the time series obtained from the EDML data behaves in a manner that differs from the behavior of the other three time series.

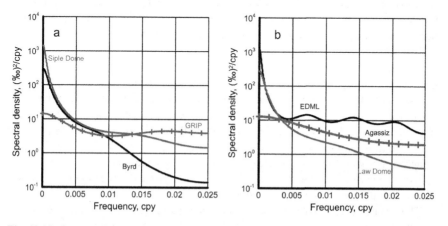

Fig. 12.10 Spectra of Antarctic ice core time series. The gray lines with symbols are the spectra of Greenland ice core time series. Note the small dynamic range of the Greenland spectra

Another important result is that the spectra diminish by orders of magnitude within the low-frequency band. In contrast to the Greenland results and to the hypothesis that climate follows a Markov model with a small regression coefficient, the behavior of climate in Antarctica if characterized with the ice core data is highly dependent upon its past values. In Greenland, this dependence is much weaker. As seen from the spectra, the temperature variations in Greenland are close to a white noise while in Antarctica the AR(1) approximation is much more persistent and its autoregressive coefficient exceeds 0.9 in three of the four cases. This phenomenon is unusual for climate variability.

Antarctica is a relatively small continent, and the four sites with the time series of $\delta^{18}O$ isotope concentrations should characterize the spatial distribution of climate properties within it. With distances between the sites in the range from 550 to 3500 km, one may expect to find some common features characterizing climate of Antarctica. To see if this happens, we will need to know statistical properties of respective bivariate time series in time and frequency domains. The shortest distance is between the Byrd and Siple Dome sites, and their altitudes differ by 900 m (Table 12.3) so one may expect at least some interdependence between the data. Respective time series are $x_{1,t}$ for Byrd and $x_{4,t}$ for Siple Dome.

The optimal time domain model selected by all five order selection criteria for this pair is **AR(3)**:

$$x_{1,t} \approx 1.34x_{1,t} + 0.07x_{4,t-1} - 0.69x_{1,t-2} - 0.03x_{4,t-2} + 0.27x_{1,t-3} + \mathbf{0.01}x_{4,t-3} + a_{1,t}$$

$$x_{4,t} \approx \mathbf{-0.06}x_{1,t-1} + 0.73x_{4,t-1} + \mathbf{0.04}x_{1,t-2} + \mathbf{0.04}x_{4,t-2} + \mathbf{0.03}x_{1,t-3} + 0.17x_{4,t-3} + a_{4,t}. \quad (12.6)$$

Remembering that the coefficients shown in bold are statistically insignificant, this bivariate model shows that the $\delta^{18}O$ and, consequently, climate variations at the sites are practically unrelated to each other. The Siple Dome values do not contain any significant contribution from the Byrd ice core while the Byrd variation seems to be weakly related to the previous value $x_{4,t-1}$ of Siple Dome's $x_{4,t}$. Similar to the Greenland case of the bivariate time series Agassiz/Renland, the lack of interdependence between the Antarctic time series situated close to each other is rather unexpected.

A close relation between two time series may also be the result of high correlation between the components of the bivariate innovation sequence $\mathbf{a}_t = [a_{1,t}, a_{4,t}]'$. This is what happens, for example, between the ENSO components SOI and NINO (see Chap. 8). However, in the current case the covariance matrix of the white noise sequence is very close to being diagonal so that the correlation coefficient between the white noise components $a_{1,t}$ (Byrd) and $a_{4,t}$ (Siple Dome) is practically zero. In agreement with this time domain model, the coherence function $\gamma_{14}(f)$ and the coherent spectra $s_{11.4}(f)$ and $s_{44.1}(f)$ between the Byrd and Siple Dome time series are statistically insignificant at all frequencies.

Thus, variations of isotope concentration presented with the ice core time series $x_{1,t}$ at Byrd and $x_{4,t}$ at Siple Dome with time scales starting roughly from 40 years and extending to millennia do not reveal any interdependence in spite of the fact that the distance between the sites is less than 600 km. In other words, climate variations at two sites situated within a relatively small area and at similar altitudes are mutually uncorrelated and reveal no common signal from outside forcing. Further analysis showed that this feature characterizes the pair EDML/Law Some as well. The physical cause for such behavior remains unclear.

12.4 Discussion and Conclusions

The basic results of our analyses given below in Table 12.6 require some clarification. The columns 5 and 8 show the innovation sequence (white noise) variances in the scalar and bivariate time domain autoregressive models, respectively, and the quantity $r(1)$ in column 6 is the first value of the time series correlation function, which in any AR(1) model coincides with the autoregressive coefficient and defines the rate of decrease of the spectral density with frequency. Our models may have higher autoregressive orders but in all cases an approximation with a Markov chain does not critically differ from the optimal model.

As follows from the table, most scalar time series are approximated with models of low order ($p = 1$ or $p = 3$). The exception is the time series from the EDML site whose spectral density behaves in a rather specific manner at frequencies above 0.005 cpy (Fig. 12.10b), possibly, due to some problems with the ice core data.

In Greenland, the time series behavior is practically controlled by the innovation sequences: their variances (column 5) are always very close to the time series variances (column 4). It means that the time series persistence is low. The ratios of variances in the bivariate case are practically the same, which means the lack of

Table 12.6 Statistical properties of scalar (AR) and bivariate (**AR**) models of isotope time series

Time series	Notation	AR order	Variance $(‰)^2$	Innovation variance $(‰)^2$	$r(1)$	AR order	Innovation variance $(‰)^2$
1	2	3	4	5	6	7	8
Greenland, 11,680 years							
Agassiz	$x_{1,t}$	1	0.14	0.11	0.43	1	0.11
Renland	$x_{2,t}$	1	0.11	0.10	0.31		0.10
GRIP	$x_{3,t}$	3	0.13	0.12	0.20	2	0.12
NGRIP	$x_{4,t}$	3	0.16	0.14	0.30		0.14
Antarctica, 9900 years							
Byrd	$x_{1,t}$	3	0.34	0.04	0.92	3	0.04
Siple Dome	$x_{4,t}$	3	0.83	0.13	0.91		0.13
EDML	$x_{2,t}$	7	0.67	0.30	0.68	3	0.31
Law Dome	$x_{3,t}$	3	0.29	0.05	0.90		0.05

correlation between the components of the bivariate system. The first values of the correlation function in column 6 are always less than 0.5 showing a weak dependence of the time series upon its past.

The situation in Antarctica is very different: the ratios of variances are much smaller—between 0.12 for Byrd and 0.45 for EDML with its doubtful reliability— and it remains the same in the bivariate cases. The persistence of the time series is much higher than at the Greenland sites, and this result is quite unexpected. According to an expert opinion by Dr. Alexei Ekaykin of the Arctic and Antarctic Research Institute in St. Petersburg, Russia, the sharp difference in the behavior of ice core time series in Greenland and Antarctica is probably the result of a strong influence of the Arctic and the North Atlantic Ocean upon Greenland's climate while variations of climate of Antarctica are much slower (personal communication). This opinion explains some of the results given in this chapter but the phenomenon needs more research based upon observations.

Our analysis of oxygen isotope $\delta^{18}O$ time series in Greenland and Antarctica produced mostly negative results. The main goals aside from methodological were to see if the time series behave similar to the behavior of climate variables and to determine whether they carry information about climate forcing at time scales from several decades to several centuries. It turned out that the time series cannot generally be treated as Gaussian and that the spatial variability of their mean values and variances is quite high even if the sites where the time series have been obtained are close to each other. Also,

- the Greenland time series are close to a Markov process with a small autoregressive coefficient, possess almost no persistence, and the dynamic range of their spectra stays within one order of magnitude.
- the scalar time series of isotope concentration in Antarctica, with the exception of the EDML data, can be approximated with autoregressive models AR(3) which do not deviate too much from Markov processes but with high AR(1) coefficients; their spectra quickly decrease with frequency and do not contain any significant peaks; these time series are persistent.

The time and frequency domain analysis of four bivariate time series of $\delta^{18}O$ showed that

- the scalar components of the time series are barely related to each other even when the distance between the sites is small (~300 km in Greenland and ~600 km in Antarctica).
- the coherence functions between the time series are low (Greenland) or statistically insignificant (Antarctica) at practically all frequencies from approximately 0.001 cpy through 0.025 cpy (time scales from 40 years to 1000 years).
- the coherent spectra are small (Greenland) or do not differ statistically significantly from zero (Antarctica).

The physical consequences of these probabilistic properties include a high spatial variability of climate properties in Greenland and in Antarctica and the lack of a common climate signal in the isotope time series. These conclusions are made under the assumption that variations of isotope concentration are related to climate variations through a linear operator, possibly but not necessarily such as a regression equation. The results are obtained for climate variability properties within the time scale interval from four or five decades to approximately a millennium.

References

Bendat J, Piersol A (2010) Random data, 4th edn. Wiley, Hoboken, New Jersey

Box G, Jenkins G, Reincell J, Ljung G (2015) Time series analysis. Forecasting and control. Wiley, Hoboken, New Jersey

Bradley R (2015) Paleoclimatology. Reconstructing climates of the quaternary. Elsevier, Amsterdam

Dobrovolski S (2000) Stochastic climate theory. Models and applications. Springer, Berlin

Gelfand I, Yaglom A (1957). Calculation of the amount of information about a random function contained in another such function. Uspekhi Matematicheskikh Nauk 12:3–52. English translation, American Mathematical Society Translation Series 2(12):199–246, 1959

Granger C, Hatanaka M (1964) Spectral analysis of economic time series. Princeton University Press, New Jersey

Hasselmann K (1976) Stochastic climate models. Part I. Theory. Tellus 28:473–485

Münch T, Laepple T (2018) What climate signal is contained in decadal- to centennial-scale isotope variations from Antarctic ice cores? Clim Past 14:2053–2070

Masson V, Vimeux F, Jouzel J et al (2000) Holocene climate variability in Antarctica based on 11 ice-core isotopic records. Quatern Res 54:348–358

Pedro, J, van Ommen T, Rasmussen A et al (2011) The last deglaciation: timing the bipolar seesaw. Clim Past 7:671–683. https://doi.org/10.5194/cp-7-671-2011

Shumway R, Stoffer D (2017) Time series analysis and its applications. Springer, Heidelberg

Vinther B, Clausen H, Fisher D et al (2008) Synchronizing ice cores from the renland and agassiz ice caps to the greenland ice core chronology. Geophys Res. https://doi.org/10.1029/2007jd009143

Vinther B, Buchardt S, Clausen H et al (2009) Holocene thinning of the Greenland ice sheet. Nat Lett. https://doi.org/10.1038/nature08355

Watanabe O, Shoji H, Satow K (2003) Dating of the Dome Fuji, Antarctica deep ice core. Mem Natl Inst Polar Res Spec Issue 57:25–37

White S, Pfister C, Mauelshagen F (2018) Palgrave handbook of climate history. Macmillan, Palgrave

Chapter 13
Application to Sunspot Numbers and Total Solar Irradiance

Abstract Time series of sunspots and total solar irradiance are studied here at monthly and annual sampling intervals as samples of scalar and bivariate (at $\Delta t = 1$ month) stationary random processes with the goal to understand their statistical predictability. Because of the high autoregressive orders, the scalar time series have to be analyzed for their properties mostly within the frequency domain. Their statistical predictability within the Kolmogorov–Wiener theory of linear extrapolation is shown to be rather strong especially for the annual data. Examples of predictions are not successful at a monthly sampling interval, probably because of the asymmetry of the solar cycle and due to our linear approach. Both SSN and TSI have probability densities that strongly differ from Gaussian, and their extrapolation through nonlinear methods may provide better results. Predicting the height of SSN peaks seems to be impossible within the KWT frame. The bivariate TSI/SSN model shows that SSN drives TSI variations; the model provides a quantitative description of the interdependences within the system. At frequencies corresponding to the solar cycle, a change of SSN by ten units causes a TSI change by about 0.05 W/m² and TSI is shown to lag behind SSN by about two months.

13.1 Introduction

The solar activity phenomena including sunspot numbers (SSN) and the total solar irradiance (TSI) have been studied in many publications dedicated to physics of the Sun and to the solar cycle; to avoid any misunderstanding, we need to define the main goals of this study, which are

- to investigate statistical properties of sunspot number variations in time and frequency domains and
- to estimate statistical properties of instrumentally measured total solar irradiation and its relations with the sunspot numbers.

In both cases, the sampling intervals are one year or one month, and we will also be interested in estimating statistical predictability of those processes and, if possible, in forecasting them.

© Springer Nature Switzerland AG 2021
V. Privalsky, *Time Series Analysis in Climatology and Related Sciences*,
Progress in Geophysics, https://doi.org/10.1007/978-3-030-58055-1_13

In a recent monograph on solar physics by Aschwanden (2019, p. 111), the success of a physically based method of solar cycle prediction is ascribed, in particular, to "a tight correlation" between a physical quantity (the solar dipole magnetic moment) and sunspot numbers, "which represents a useful quantitative relationship between an observable … and a physical parameter". This means an understanding of the probabilistic nature of the solar cycle, which can be explained as a physical process and predicted through, possibly, a physical model and through statistical information about the behavior of sunspot numbers. For our purposes, the probabilistic nature of sunspot variability is clear for both physical reasons (sunspots are controlled by the Sun's magnetic activity) and because of the necessity to apply probabilistic methods to physical processes—solar cycle and sunspot number variations—that cannot be regarded as deterministic and predicted without an error.

The issue of SSN prediction is not new in solar sciences. Examples are numerous (e.g., Li et al. 2002; Aguirre et al. 2008; Gkana and Zachilas 2015; Liu et al. 2019) but, to the best of the author's knowledge, they all contain the same essential flaw that exists in Earth sciences: none of those publications ever uses or even mentions the classical Kolmogorov–Wiener theory of extrapolation (KWT) that provides a solution for predictions of stationary random processes. Why the theory remains unknown to climatologists and solar scientists remains a puzzle since many decades ago.

The sunspot variability is known from observations for several centuries, and the behavior of the record gives no ground for not recognizing the generating process as stationary. This statement is supported by a simple physical consideration: if this random process was nonstationary, it would mean that we have been and still are observing an evolution of Sun's magnetic activity at time scales of decades or a few centuries, which seems to be too fast for such a large celestial body. We will study properties of the sunspot time series and attempt to extrapolate it at annual and monthly time scales.

The approach to the task of SSN prediction as a stationary random process is also not new. Thus, in a recent publication on the subject by Abdel-Rahman and Mourzuk (2018), sunspots are extrapolated as an autoregressive sequence of order one, that is, as a Markov chain. Such approximation for the time series of monthly SSN is in a stark conflict with observations because it is much more complicated stochastic model; yet, a Markov chain is a stationary random process. Thus, the sunspot time series should and will be regarded here as a sample record of a stationary random process. The stationarity hypothesis agrees with observation of both SSN and TSI.

As the KWT-based methods produce results with a minimal possible variance of linear prediction error (or absolutely minimal possible variance of prediction error in the Gaussian case), any linear method of prediction not based upon the theory must be treated as incorrect. It will be shown below that the task of extrapolating the SSN and TSI time series will probably require a nonlinear approach but in this book, we will treat only the linear case.

The observed total solar irradiance is closely connected to sunspot numbers; this property is used for TSI reconstructions into the past but we will limit ourselves with studying properties of the observed time series, including its statistical predictability and extrapolating TSI and SSN at the annual and monthly time scales.

A possible influence of changes in solar irradiance upon the Earth climate has been and still is a subject for discussions in climatology and in solar research in spite of the numerous unsuccessful attempts to prove its existence. A thorough review is given by Gray et al. (2010). The TSI variability as a possible external forcing of climate is often studied in connection to variations of sunspots numbers (e.g., Egorova et al. 2018; Fröhlich 2019). In this chapter, we will apply the tools of time series analysis to describe statistical properties of these basic time series that are used, in particular, in attempts to statistically prove the effect of the very delicate variability of solar radiation upon climate. The goal here is to show that properties of the time series of SSN and TSI, including their interdependence, are quite complicated in both time and frequency domains and must be taken into account in any attempt to prove TSI's effect upon climate, to restore the total solar irradiance by using sunspot numbers as a proxy or for other research involving SSN and TSI time series as sample records of random processes. The issue of time series reconstruction will not be discussed because the author's method of reconstruction based upon theory of random processes has been suggested in a recent publication (Privalsky 2018); its efficiency is verified in Chap. 9.

The properties of SSN and TSI will also be studied here for the case when these time series are regarded as components of a bivariate time series TSI/SSN.

13.2 Properties of Sunspot Number Time Series

The monthly SSN data are produced by the WDS-SILSO, Royal Observatory of Belgium, Brussels, Belgium, and can be found at the site http://sidc.oma.be/silso/ DATA/SN_m_tot_V2.0.txt. The sunspot variations can be studied at different time scales and in different configurations. In this chapter, we will be dealing only with the time series of sunspots, not the group numbers. The data source is https://cli mexp.knmi.nl/data/isunspots.dat. The monthly TSI values have been taken from the site https://climexp.knmi.nl/data/itsi.dat. Having in mind the subsequent analysis of TSI and SSN as a bivariate system, the TSI and SSN time series will be notated as $x_{1,t}$ and $x_{2,t}$.

The annual SSN data $x_{2,t}$ from 1749 through 2019 ($N = 271$) are shown in Fig. 13.1a. As seen from the graph, the time series can be regarded as a sample of a random process containing strong cyclicity with low-frequency variations. It does not contain any trend, and it successfully passes the test for stationarity suggested in Chap. 4. In particular, the confidence intervals for variance estimates obtained for the entire series and for its halves overlap, and the spectral density estimates are very close to each other (Fig. 13.1b). The spectra of the entire time series and its halves contain a sharp peak at $f \approx 0.095$ cpy, which is close to the time scale of the 11-year cycle.

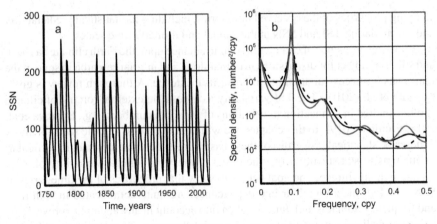

Fig. 13.1 Time series of annual SSN, 1749-2019 (**a**) and autoregressive spectral estimates for the entire time series (solid black line) and its halves (**b**)

The optimal autoregressive (AR) model for the time series of annual SSN selected by all five order selection criteria is AR(9), that is,

$$x_{2,t} = \sum_{k=1}^{9} \varphi_k x_{2,t-k} + a_{2,t},$$

where φ_k are the autoregressive coefficients and $a_{2,t}$ is the innovation sequence. This model is too cumbersome, and we will have to waive a detailed time domain analysis. The SSN variance $\sigma_2^2 \approx 4000$ number^2 while the variance of its innovation sequence $a_{2,t}$ is 614 number^2; thus, the statistical predictability of this time series is rather high: the criterion $r_e(1) \approx 0.92$ (see Chap. 6). This happens due to the cyclic character of SSN variations.

The predictability of the annual SSN data as a stationary random process is illustrated with the graphs of predictability criteria (Fig. 13.2) and with extrapolation results shown in Fig. 13.3.

The prediction quality criterion quickly drops to about 0.7 and stays above 0.6 at the lead time $\tau = 12$ years while the relative predictability criterion $\rho(\tau) = \sqrt{1 - r_e^2(\tau)}$ achieves approximately 0.7 and 0.9 at $\tau = 3$ years and $\tau = 18$ years, respectively. The prediction quality criterion coincides with the correlation coefficient between the unknown true future value of the time series and its linear extrapolation at a lead time τ. This moderately high statistical predictability of SSN happens because of its cyclic behavior.

A linear extrapolation of SSN during the 24th solar cycle starting from year 2010 is successful, and the predicted values of SSN lie close to subsequent observations (crosses in Fig. 13.3a). At the same time, the relative predictability criterion $\rho(\tau)$ increases to 0.6 at $\tau = 2$ years meaning that even at this small lead time the confidence interval for the prediction curve becomes equal to $\pm\sigma_2$; it grows by about 20% at

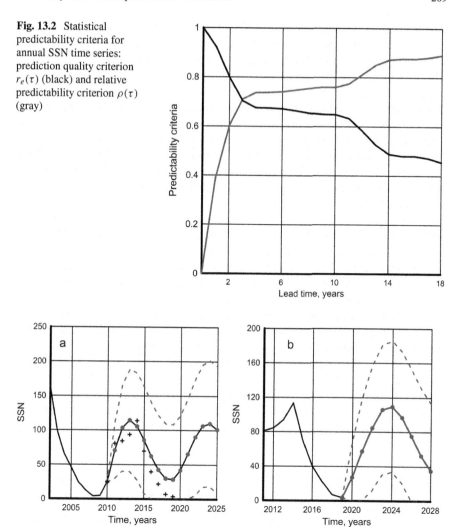

Fig. 13.2 Statistical predictability criteria for annual SSN time series: prediction quality criterion $r_e(\tau)$ (black) and relative predictability criterion $\rho(\tau)$ (gray)

Fig. 13.3 Forecasting annual values of SSN from 2010 (**a**) and from 2019 (**b**). The crosses show the observed values.

$\tau = 10$ years. On the whole, one may say that the forecasts of annual SSN time series will probably stay more or less reliable at lead times up to about 3–5 years. The prediction of annual SSN starting from 2019 is shown in Fig. 13.3b. Note that the confidence interval for it is growing pretty fast with the lead time τ.

For the purposes of solar physics, the rather accurate prediction of SSN in Fig. 13.3a may not be regarded as an achievement. According to Svalgaard et al. (2005), a key goal of solar-terrestrial physics is to predict the peak amplitude of the solar cycle. This means that solar physics needs predictions of random events while the Kolmogorov–Wiener theory predicts random processes. As the process is stationary, the predicted trajectory will tend to the mean value of the process and, consequently, in this specific case with a cyclic random process, the amplitude of any predicted cycle will probably be smaller than the amplitude of the previous cycle. The theory is not built to predict individual events.

At the same time, there can be no other linear statistical method of extrapolation that would have a smaller error variance. The time series consisting of annual values of SSN is not Gaussian so that that its nonlinear extrapolation may be more accurate. Applying any method of mathematical statistics for forecasting individual events such as the amplitude of a cycle would be improper because in a time series (SSN in this case) any future value depends upon the past behavior, and this property means that the only solution would have to be based upon a theory of extrapolation of random processes, linear or nonlinear.

Several statistical approaches can be tested for predicting the amplitude of the next solar cycle. One is to try finding an operator that would minimize the error variance of the amplitude prediction assuming that all solar cycles have the same period. But actually, the period of the cycle is not constant. Another approach would be to analyze the sequence of peak values as a time series. Our analysis shows that if the period is assumed to be constant, the sequence of peak values behaves as a white noise so that the future values are not predictable. A similar result is obtained for the sequence of cycle lengths. It looks like obtaining reliable results in forecasting the amplitude of the next solar cycle using a probabilistic approach will not be easy.

The SSN time series at a monthly sampling interval from January 1749 through December 2019 contains $N = 3252$ values. Its graph is shown in Fig. 13.4a with respective spectral estimates in Fig. 13.4b.

This time series passes the test for stationarity, and its best autoregressive approximation is an AR(34) model. Due to the high order, it cannot be analyzed in the time

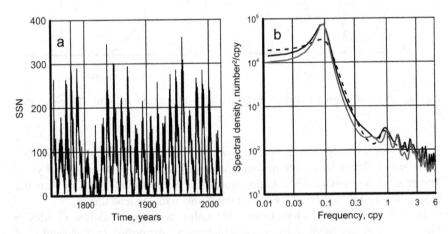

Fig. 13.4 Time series of monthly SSN, 1749-2019 (**a**) and autoregressive spectral estimates for the entire time series (solid black line) and its halves (**b**)

domain. However, we can study its statistical predictability within the Kolmogorov–Wiener theory. As seen from Fig. 13.5, the statistical predictability of the monthly SSN time series is rather high: the 90% confidence interval for the forecast curve becomes equal to $\pm\sigma_2$ at $\tau = 12$ months.

Formally, the forecasts shown in Fig. 13.6 are successful because they satisfy the requirement that most observed values should be within the confidence interval; however, the forecast in Fig. 13.6b shows a definite positive bias. The reason for this is probably the asymmetry of the solar cycle's shape at the monthly rate: the values

Fig. 13.5 Statistical predictability criteria for monthly SSN time series: prediction quality criterion $r_e(\tau)$ (black) and relative predictability criterion $\rho(\tau)$ (gray)

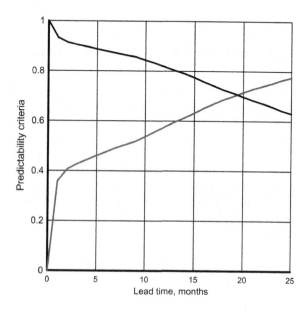

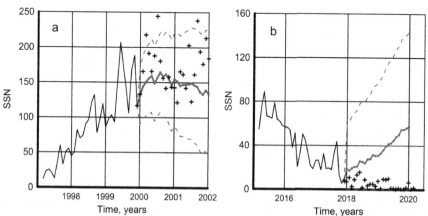

Fig. 13.6 Forecasts of monthly SNN time series from December 1999 (**a**) and from December 2017 (**b**)

of SSN have a tendency to stay longer at the cycle's minimum stage as happened, for example, between 1887 and 1891 and between 1911 and 1915; it remains low since the late 2017. Obviously, the cycle asymmetry cannot be taken into account by the extrapolation algorithm given with Eq. (6.11). The forecasts in Fig. 13.6b stay within the confidence interval at lead times over 20 months but seemingly if the starting points of extrapolation lie close to an early part of the cycle's minimal stage one may expect a lasting positive bias. The asymmetry property of solar cycles may be another (along with the non-Gaussianity) reason to try nonlinear methods for its extrapolation, in particular, the asymmetric parametric models (De Gooijer 2017).

13.3 Properties of Total Solar Irradiance Time Series

The time series of monthly TSI values is available from February 1978 through April 2018 (see https://climexp.knmi.nl/data/itsi.dat). An up-to-date time series does not seem to be available. The time series consisting of annual values calculated for each year from February through January is shown in Fig. 13.7a, and its probability density function is non-Gaussian. It behaves in a manner that allows one to regard it as a sample of a stationary process. The length of this time series is 40 and it can be treated a Gaussian.

The optimal autoregressive order for the TSI time $x_{1,t}$ at $\Delta t = 1$ year is $p = 3$ and it has been indicated by all order selection criteria. The time domain model of TSI is

$$x_{1,t} \approx 1.02x_{1,t-1} - \mathbf{0.10}x_{1,t-2} - 0.42x_{1,t-3} + a_{1,t}. \qquad (13.1)$$

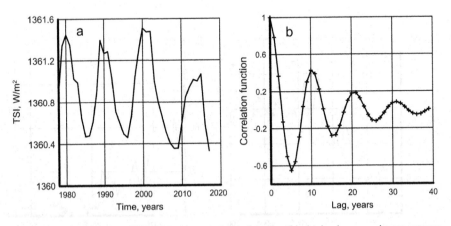

Fig. 13.7 Annual values of TSI (**a**) and its correlation function (**b**) obtained as a maximum entropy extension of the first three sampling estimates

The second AR coefficient does not differ from zero at a confidence level 0.9. The natural frequency corresponding to Eq. (13.1) is 0.096 cpy which gives the period of TSI's cyclic variations as 10.3 years. The 11-year period lies inside the 90% confidence limits for this estimate.

As shown in Chap. 4, the AR model given in this case with Eq. (13.1) provides an expression for the correlation function of the TSI time series, which is obtained as the maximum entropy extension of the first p sample values of the correlation function. The extension is given in Fig. 13.7b, and it shows that the model (13.1) has detected the cycling character of this stationary random process.

The total variance $\sigma_1^2 \approx 0.132$ W/m^2 and the innovation sequence variance $\sigma_a^2 \approx$ 0.026 W/m^2 so that the prediction quality criterion $r_e(1) = \sqrt{1 - \sigma_a^2/\sigma_1^2}$ is close to 0.9 meaning that the annual variations of total solar radiance possess high statistical predictability. However, such predictions at $\Delta t = 1$ year with a very short time series may not be reliable.

Multiplying Eq. (13.1) by $x_{1,t}$ and finding the mathematical expectation leads to the equation

$$\sigma_1^2 \approx 1.02R(1) - 0.10R(2) - 0.42R(3) + \sigma_a^2,$$

where $R(k)$ is the covariance function of $x_{1,t}$. Calculations show that the variance σ_1^2 of the annual TSI time series consists of contributions from the previous value $x_{1,t-1}$ (close to 80%) and the innovation sequence $a_{1,t}$ (20%). The role of the previous values $x_{1,t-2}$ and $x_{1,t-3}$ is relatively small because of the fast decrease in the correlation function.

The cyclic property of annual TSI is also seen from its estimates of spectral density (Fig. 13.8). Though the TSI time series is very short, both the autoregressive (or MEM) and Thomson's (MTM) estimates show a strong signal at the frequency of about 0.096 cpy (close to the time scale of 11 years). The AR estimate seems to be

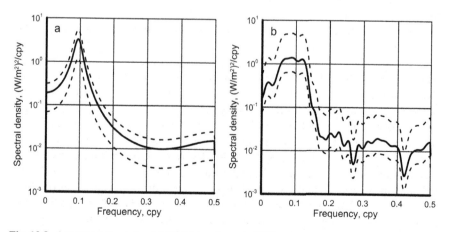

Fig. 13.8 Autoregressive (**a**) and MTM (**b**) estimates of TSI spectrum

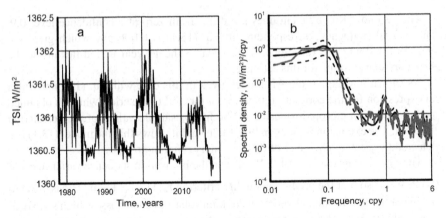

Fig. 13.9 Time series of monthly TSI (**a**) and its autoregressive (**a**) and MTM (**b**) estimates

better: it has a smooth shape without any random deviations, and its 90% confidence interval is tighter.

At the sampling interval of one year, the TSI time series is too short to study it as a component of the bivariate system TSI/SSN.

Consider now the TSI data as a scalar sample record of length $N = 483$ months (Fig. 13.9a). This time series also shows an eleven years cycle but, in contrast to the annual case shown in Fig. 13.7, the graph contains strong high-frequency variations. The optimal AR order for this time series is $p = 20$ and the maximum entropy extension of the correlation function corresponding to this model does not show a clear cyclic behavior. The spectral maximum occurs exactly at the frequency $f = 0.09$ cpy, that is, at the time scale of the 11-year cycle (Fig. 13.9b). However, the maximum is not expressed as clearly as in the case of annual data.

The variance σ_1^2 of the TSI time series $x_{1,t}$ at the monthly sampling interval is 0.170 (W/m^2)^2 while the variance of innovation sequence $a_{1,t}$ is 0.043 (W/m^2)^2; thus, the prediction quality criterion $r_e(1) \approx 0.86$.

The relative predictability criterion $\rho(\tau) = \sigma_\varepsilon(\tau)/\sigma_1$ shows the growth of prediction error from 0.5 at the unit (one month) lead time to about 0.9 at $\tau = 25$ months (Fig. 13.10). The 90% confidence interval becomes equal to $\pm\sigma_1$ at $\tau = 5$ months and at $\tau = 7$ months for the initial extrapolation dates in 2016 and 2018, respectively. This means that an interval of 5–7 months long can be regarded as the limit of statistical predictability for the TSI time series.

The results of extrapolation are satisfactory for the first several months after September 2016 and remain within the 90% confidence limits for longer lead times (Fig. 13.11a). Still, the time series of monthly TSI values is much less regular (cyclic) than the time series of annual TSIs, and unsuccessful results are more probable for it than in the case of annual TSI data. One may expect the observed values of TSI to stay within the 90% confidence limits at least up to the end of 2018. The later data are not available. Judging by the behavior of monthly TSI, the nonlinear approach to its extrapolation may produce better results.

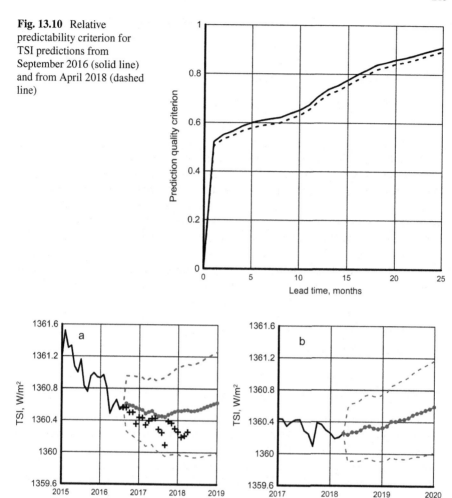

Fig. 13.10 Relative predictability criterion for TSI predictions from September 2016 (solid line) and from April 2018 (dashed line)

Fig. 13.11 Extrapolation of monthly TSI (gray) starting from September 2016 (**a**) and from April 2018 (**b**). The observed values are shown with black crosses.

The bivariate time series with the scalar components TSI and SSN at a monthly sampling interval is analyzed here assuming that TSI and SSN present the output and input of a linear system TSI/SSN. It has been studied in the time domain and, partially, in the frequency domain in Privalsky (2018) but a full description in the frequency domain has not been discussed before. For convenience, the values of SSN have been divided by 10.

The optimal model for the bivariate time series TSI/SSN at the monthly sampling interval is the following **AR**(3) sequence:

$$x_{1,t} \approx 0.37x_{1,t-1} + 0.03x_{2,t-1} + 0.16x_{1,t-2} + 0.08x_{1,t-3} - 0.01x_{2,t-3} + a_{1,t}$$
$$x_{2,t} \approx 0.57x_{2,t-1} + 0.13x_{2,t-2} + 1.94x_{1,t-3} + 0.18x_{2,t-3} + a_{2,t} \qquad (13.2)$$

and the covariance matrix of innovation sequence is

$$\mathbf{P_a} \approx \begin{bmatrix} 0.036 & -0.151 \\ -0.151 & 6.02 \end{bmatrix}. \qquad (13.3)$$

The statistically insignificant coefficients are not shown in Eq. (13.2).

According to Eq. (13.2), the TSI ($x_{1,t}$) behavior depends upon three of its past values and upon values of SSN that occurred one and three months ago. The small values of respective AR coefficient at $x_{2,t-k}$ should not bother us because they are statistically significant and because after dividing the original values of SSN by 10 its variance $\sigma_2^2 \approx 52.2$ is still more than two orders of magnitude larger than σ_1^2.

The SSN value $x_{2,t}$ depends upon its past for three months and upon the TSI value three months ago while SSN remembers its three preceding values. The components of the TSI and SSN variances can be determined by using Eqs. (7.12) and (7.13):

$$\sigma_1^2 = \sum_{k=1}^{p} [\phi_{11}^{(j)} R_{11}(k) + \phi_{12}^{(j)} R_{21}(k)] + \rho_{12}^2 P_{11} + (1 - \rho_{12}^2) P_{11}$$

$$\sigma_2^2 = \sum_{k=1}^{p} [\phi_{21}^{(j)} R_{21}(k) + \phi_{22}^{(j)} R_{22}(k)] + \rho_{12}^2 P_{22} + (1 - \rho_{12}^2) P_{22}.$$

Here, $R_{ij}(k)$ are the elements of the covariance matrix $\mathbf{R}_{12}(k)$ of the time series $\mathbf{x}_t = [x_{1,t}, x_{2,t}]'$. The results of calculations are shown in Table 13.1. Almost one half of TSI variance is contributed by TSI past values and almost one-third, by the past values of SSN. The rest is provided by the innovation sequence. The SSN variance is barely dependent upon TSI because most of the variance is generated by the past values of SSN. This dependence results from a feedback between SSN and TSI due to the presence of a TSI term $x_{1,t-3}$ in the second equation of the model (13.2).

Consider now the structure of the TSI/SSN system in the frequency domain. The estimated spectra are shown in Figs. 13.9b and 13.4b so we will continue with the coherence function and the gain factor which can be regarded as frequency-dependent equivalents of correlation and regression coefficients. These quantities are shown in

Table 13.1 Summands of TSI and SSN variances

Source	TSI (%)	SSN (%)
From TSI	48	8
From SSN	31	81
From $a_{1,t}$	2	1
From $a_{2,t}$	19	10

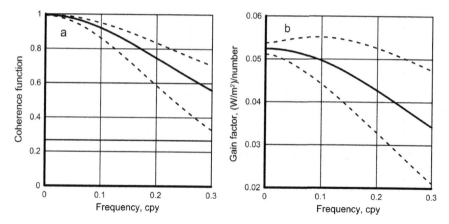

Fig. 13.12 Coherence function (**a**) and gain factor (**b**) between TSI and SSN. The horizontal line above 0.2 in the coherence graph shows the 90% upper confidence limit for the true zero coherence

Fig. 13.12 for the frequency band up to 0.3 cpy; at higher frequencies, the values of the spectral density are very low.

The coherence stays statistically significant within the entire frequency band and amounts to 0.93 at 0.09 cpy (the 11-year cycle). The gain factor shows how variations of SSN are transformed into variations of TSI; in our case, a change of SSN by ten units (of the original SSN, not divided by 10) causes a change of about 0.050 W/m² in TSI.

Incidentally, if the regression equation is used for reconstructing past values of TSI, the regression coefficient equals 0.043 W/m² per ten units of SSN. This negative bias error occurs because the use of linear regression is equivalent to averaging the gain factor over the entire frequency band; it is one of the reasons why the regression approach should not be used with time series whose properties, including the gain factor, practically always vary with frequency.

A positive phase factor between TSI and SSN (Fig. 13.13a) shows that TSI lags behind SSN as it should be between the output and input of a physical system. Thus, the assumption of SSN being the input and TSI the output was correct. The time lag at 0.09 cpy is about 0.17 year or about two months (Fig. 13.13b).

Hopefully, the description of properties of SSN and TSI time series given here may be useful for a better understanding of these processes including the physical mechanism responsible for generation of solar radiation.

As mentioned before, the TSI and SSN time series have non-Gaussian probability densities, which means that better forecasting results can only be obtained with nonlinear methods of extrapolation. Finding an efficient approximation for the extrapolation function will be one of the difficulties in switching to a nonlinear approach. The problem of cycle asymmetry seems to be important for both SSN and TSI at a monthly sampling interval. Several types of nonlinear transformations of time series are given in the book by Bendat (1990), and the subject of nonlinear extrapolation is discussed in a recent book by De Gooijer (2017).

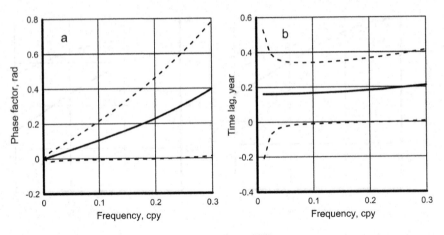

Fig. 13.13 Phase factor and time lag between TSI and SSN

In conclusion, consider the relations between TSI and the monthly surface global temperature by analyzing the time series in accordance with theory of multivariate random processes. It can be assumed in advance that the failures to find such dependence are probably related to the extremely small variability of TSI: its coefficient of variation $c_1 = \sigma_1/\bar{x}_1 \approx 3 \times 10^{-4}$, where σ_1 and $\bar{x}_1$ are the root means square and mean values of TSI ($x_{1,t}$). Nevertheless, we will analyze the bivariate time series containing TSI as the input and monthly global temperature $x_{3,t}$ as the output. The time interval is from February 1978 through April 2018, and the global temperature data are taken from the site https://crudata.uea.ac.uk/cru/data/temperature. The time series are shown in Fig. 13.14; its length is 483 months. The trend in the temperature time series $x_{3,t}$ has not been deleted but its deletion does not lead to noticeable changes in the results of analysis given below.

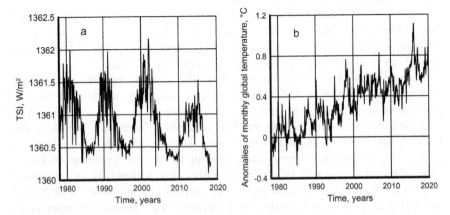

Fig. 13.14 Monthly total solar irradiance (**a**) and global temperature (**b**), February 1978–April 2018

The optimal autoregressive model for the bivariate time series $\mathbf{x}_t = [x_{3,t}, x_{1,t}]'$ is AR(3):

$$x_{3,t} \approx 0.50x_{3,t-1} + 0.04x_{1,t-1} + 0.24x_{3,t-2} + a_{3,t}$$
$$x_{1,t} \approx 0.46x_{1,t-1} + 0.20x_{1,t-2} + 0.25x_{1,t-3} + a_{1,t}. \qquad (13.4)$$

According to this equation, the global temperature depends upon its two previous values and upon the value of TSI observed one month ago. The autoregressive coefficient $\varphi_{13}^{(1)} \approx 0.04$ is statistically significant even at a significance level 0.05, and it is the only quantity that connects the global temperature $x_{3,t}$ to the solar irradiance $x_{1,t}$ because the innovation sequences $a_{3,t}$ and $a_{1,t}$ are not correlated with each other. Yet, some influence of irradiance upon temperature is present in the first equation of the system (13.4) so we need to see the cross-correlation and coherence function between the two time series.

The maximum value of the cross-correlation function barely exceeds 0.2 (Fig. 13.15a) while the coherence function quickly decreases from about 0.4 to 0.27—the upper 90% confidence limit for the true zero coherence shown with a gray line (Fig. 13.15b). The frequency axis in Fig. 13.15b is given in a logarithmic scale to reveal the several frequencies where the lower confidence limit for the coherence estimate exceeds zero. The reliability of respective estimates of coherence is low because those frequencies correspond to time scales comparable with the time series length. Most values of the coherence function (actually, 12 out of 500) have zero as its lower 90% confidence level.

The only conclusion that can be drawn from this experiment is that the influence of solar irradiation variations upon the global temperature lies below the level of detectability.

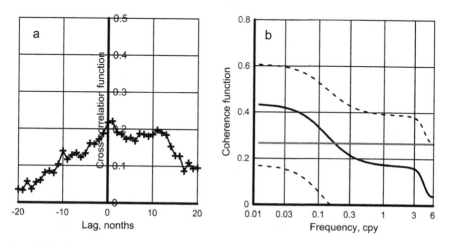

Fig. 13.15 Estimates cross-correlation (**a**) and coherence (**b**) functions between solar irradiance and global temperature

References

Abdel-Rahman H, Mourzuk B (2018) Statistical method to predict the sunspots number. NRIAG J Astron Geophys 7:175–179

Aguirre L, Letellie C, Maquet J (2008) Forecasting the time series of sunspot numbers. Solar Phys 249:103–120

Aschwanden M (2019) New millennium solar physics. Springer Nature, Switzerland

Bendat J (1990) Nonlinear system analysis and identification from random data. Wiley, New York

De Gooijer J (2017) Elements of nonlinear time series analysis and forecasting. Springer, Amsterdam

Egorova T, Schmutz W, Rozanov E et al (2018) Revised historical solar irradiance forcing. A&A, A85

Fröhlich C (2019) How variable is the solar constant. https://www.hfsjg.ch/wordpress/workshops/2019_EPS/Froehlich_Claus.pdf

Gkana A, Zachilas L (2015) Sunspot numbers: data analysis, predictions and economic impacts. J Eng Sci Technol Rev 8:79–85

Gray L, Beer J, Geller M et al (2010) Solar influences on Climate. Rev Geophys 48:RG4001

Li K, Zhan L, Wang J et al (2002) A method for the prediction of relative sunspot number for the remainder of a progressing cycle with application to cycle 23. A&A 302:301–307

Liu J, Zhao J, Lin H (2019) Prediction of the sunspot number with a new model based on the revised data. Solar Phys 294:157

Privalsky V (2018) A new method for reconstruction of solar irradiance. J Atmos Sol-Ter Phy 72:138–142

Svalgaard L, Cliver E, Kamide Y (2005) Sunspot cycle 24: smallest cycle in 100 years? Geophys Res Lett 32. https://doi.org/10.1029/2004gl021664

Chapter 14
Multivariate Time and Frequency Domain Analysis

Abstract The time series is called multivariate here if the number M of its scalar components exceeds two. Multivariate models describe linear systems with one output and $M - 1$ inputs. Studying such series in time domain can be cumbersome, and the accent moves to frequency domain analysis based upon the $M \times M$ spectral matrix obtained through Fourier transform of the time domain model. For $M > 2$, the number of quantities that characterize the system increases due to the appearance of several new quantities: partial coherence functions and coherent spectra, gain and phase factors for every tract of the system plus additional spectral densities, one multiple coherence, and one multiple coherent spectrum. These functions produce frequency domain descriptions of each tract of the system and of the net effect of all input processes characterized with multiple coherences and spectra. The analysis takes into account linear relations between the input time series. An M-variate model $\mathbf{AR}(p)$ contains $M^2 p$ autoregressive coefficients. An example of time and frequency domain analysis of a simulated trivariate time series shows how to estimate the time series properties and interpret the results. Analysis of global, oceanic, and terrestrial surface temperature data as a two-input system reveals a pazzling change in system's properties.

The goal of this chapter is to describe the process of analysis of linear systems with one output process and more than one input. The examples include studying a simulated trivariate time series and two trivariate climatic time series. They present a system with the global temperature as the output and hemispheric time series as inputs and a system with the global time series as the output and oceanic and terrestrial data as inputs.

The analysis is conducted in both time and frequency domains with an accent upon frequency-dependent characteristics. The frequency domain part is based upon Chap. 7 of the book by Bendat and Piersol (2010).

The time series dimension here is $M = 3$ and the random process which is to be analyzed is $\mathbf{x}_t = [x_{1,t}, x_{2,t}, x_{3,t}]'$; the sampling rate $\Delta t = 1$. With $M = 3$, the bivariate autoregressive stochastic difference Eq. (7.4)

© Springer Nature Switzerland AG 2021
V. Privalsky, *Time Series Analysis in Climatology and Related Sciences*,
Progress in Geophysics, https://doi.org/10.1007/978-3-030-58055-1_14

$$x_{1,t} = \varphi_{11}^{(1)} x_{1,t-1} + \varphi_{12}^{(1)} x_{2,t-1} + \cdots + \varphi_{11}^{(p)} x_{1,t-p} + \varphi_{12}^{(p)} x_{2,t-p} + a_{1,t}$$

$$x_{2,t} = \varphi_{21}^{(1)} x_{1,t-1} + \varphi_{22}^{(1)} x_{2,t-1} + \cdots + \varphi_{21}^{(p)} x_{1,t-p} + \varphi_{22}^{(p)} x_{2,t-p} + a_{2,t}$$

has one more component so that

$$
\begin{aligned}
x_{1,t} &= \varphi_{11}^{(1)} x_{1,t-1} + \varphi_{12}^{(1)} x_{2,t-1} + \varphi_{13}^{(1)} x_{3,t-1} + \cdots + \varphi_{11}^{(p)} x_{1,t-p} \\
&\quad + \varphi_{12}^{(p)} x_{2,t-p} + \varphi_{13}^{(p)} x_{3,t-p} + a_{1,t} \\
x_{2,t} &= \varphi_{21}^{(1)} x_{1,t-1} + \varphi_{22}^{(1)} x_{2,t-1} + \varphi_{23}^{(1)} x_{3,t-p} + \cdots + \varphi_{21}^{(p)} x_{1,t-p} \\
&\quad + \varphi_{22}^{(p)} x_{2,t-p} + \varphi_{23}^{(p)} x_{3,t-p} + a_{2,t} \\
x_{3,t} &= \varphi_{31}^{(1)} x_{1,t-1} + \varphi_{32}^{(1)} x_{3,t-1} + \varphi_{33}^{(1)} x_{3,t-p} + \cdots + \varphi_{31}^{(p)} x_{1,t-p} \\
&\quad + \varphi_{32}^{(p)} x_{2,t-p} + \varphi_{33}^{(p)} x_{3,t-p} + a_{3,t}.
\end{aligned}
\tag{14.1}
$$

The (2×2) matrix of autoregressive coefficients

$$\Phi_j = \begin{bmatrix} \varphi_{11}^{(j)} & \varphi_{12}^{(j)} \\ \varphi_{21}^{(j)} & \varphi_{22}^{(j)} \end{bmatrix}$$

becomes

$$\Phi_j = \begin{bmatrix} \varphi_{11}^{(j)} & \varphi_{12}^{(j)} & \varphi_{13}^{(j)} \\ \varphi_{21}^{(j)} & \varphi_{22}^{(j)} & \varphi_{23}^{(j)} \\ \varphi_{31}^{(j)} & \varphi_{32}^{(j)} & \varphi_{33}^{(j)} \end{bmatrix}, \quad j = 1, \ldots, p.$$

The bivariate innovation $\mathbf{a}_t = [a_{1,t}, a_{2,t}]'$ is now $\mathbf{a}_t = [a_{1,t}, a_{2,t}, a_{3,t}]'$, and the (2×2) covariance matrix

$$\mathbf{P_a} = \begin{bmatrix} P_{11} & P_{12} \\ P_{21} & P_{22} \end{bmatrix}$$

becomes

$$\mathbf{P_a} = \begin{bmatrix} P_{11} & P_{12} & P_{13} \\ P_{21} & P_{22} & P_{23} \\ P_{31} & P_{32} & P_{33} \end{bmatrix}.$$

Thus, the trivariate stochastic difference Eq. (14.1) describes a linear system with the output $x_{1,t}$ and the inputs $x_{2,t}$, $x_{3,t}$. It takes into account possible dependence of each of the scalar components upon its own past values and upon past values of the other two components including the dependence between the input time series.

An optimal model $\mathbf{AR}(p)$ for a trivariate time series should be selected with the help of order selection criteria. Choosing the order arbitrarily is improper.

14.1 Time Domain Analysis

The time domain analysis within the framework of autoregressive modeling is essentially the same as in the bivariate case, and most computations can be done with the VAR package in Shumway and Stoffer (2017) and in MATLAB. The time domain information should include

- covariance and correlation matrices $\mathbf{R}(k)$ and $\mathbf{r}(k)$, $|k| = 0, \ldots, K$, where the maximum lag K should not exceed approximately one-tenth of the time series length N,
- values of order selection criteria to determine an optimal order $\mathbf{AR}(p)$,
- estimated matrix autoregressive coefficients for each of the three components of the $\mathbf{AR}(p)$ model,
- error variances or RMS errors for estimated AR coefficients, and
- estimated covariance and correlation matrices $\mathbf{P_a}$ and $\boldsymbol{\rho_a}$ of the innovation sequence $\mathbf{a}_t$ with respective error variances.

The diagonal elements of the matrix $\mathbf{P_a}$ are the variances of white noise sequences $a_{1,t}$, $a_{2,t}$, and $a_{3,t}$, which constitute the unpredictable elements of the system (14.1). The correlation matrix $\boldsymbol{\rho_a}$ defines the cross-correlation coefficients between the innovation sequences. It is obtained by normalizing the elements of $\mathbf{P_a}$ by respective variances and covariances of innovation sequence components. The estimate of $\mathbf{R}(k)$ can also be used for determining the share of the components' variances similar to how it is done in the scalar and bivariate cases [Eqs. (3.21), (7.12), and (7.13)].

As in the scalar and bivariate cases discussed in Chap. 3 and 7, the trivariate time domain analysis allows one to estimate the predictability of time series components, which is done by comparing variances of the components with the variances of respective innovation sequences. The time domain analysis also provides information sufficient for deciding whether the system contains any Granger causality and feedback.

The stochastic difference equations (Eq. 14.1) can be used in studies of multivariate teleconnections and in time series reconstructions with more than one proxy time series. It also allows one to extrapolate time series which have more than one time-dependent predictor.

14.2 Frequency Domain Analysis

The multivariate frequency domain analysis begins with the spectral matrix. When $M = 3$, the spectral matrix

$$\mathbf{s}(f) = \begin{bmatrix} s_{11}(f)\, s_{12}(f)\, s_{13}(f) \\ s_{21}(f)\, s_{22}(f)\, s_{23}(f) \\ s_{31}(f)\, s_{32}(f)\, s_{33}(f) \end{bmatrix}. \tag{14.2}$$

Here, the diagonal elements present spectral densities of the output and input processes while the off-diagonal elements $s_{12}(f)$ and $s_{13}(f)$ are the complex-valued cross-spectra between the output and inputs; the quantity $s_{23}(f)$ is the cross-spectrum between the inputs. The spectral matrix is Hermitian so that $s_{ij}(f) = \bar{s}_{ji}(f)$, where the bar means complex conjugation. The matrix is obtained through a Fourier transform of the autoregressive Eq. (14.1), and all functions that define the behavior of a trivariate time series in the frequency domain are calculated through its elements. In addition to the spectra $s_{11}(f)$, $s_{22}(f)$, and $s_{33}(f)$, they include

- three coherence functions:
    - multiple coherence $\gamma_{1:23}(f)$,
    - partial coherences $\gamma_{12.3}(f)$, and $\gamma_{13.2}(f)$,
- three coherent spectra:
    - multiple coherent spectrum $s_{1:23}(f) = \gamma_{1:23}^2(f)s_{11}(f)$,
    - partial coherent spectra $s_{12.3}(f)$, and $s_{13.2}(f)$,
- gain factors $g_{12.3}(f)$, $g_{13.2}(f)$, and
- phase factors $\varphi_{12.3}(f)$, $\varphi_{13.2}(f)$.

The gain and phase factors present the real and imaginary parts of the complex-valued frequency response function connecting the output processes to the inputs.

All these functions take into account possible linear relations between the input processes $x_{2,t}$ and $x_{3,t}$. By themselves, the off-diagonal elements of the spectral matrix (14.2) explain the frequency domain properties of the time series $\mathbf{x}_t$ only in the case when the inputs are not related to each other, that is, when $\gamma_{23}(f) = 0$. Therefore, neglecting possible linear relations between the inputs is improper, and the ordinary coherence function $\gamma_{12}(f)$ and $\gamma_{13}(f)$ cannot serve as substitutes for partial coherences $\gamma_{12.3}(f)$ and $\gamma_{13.2}(f)$. In other words, when the dimension of the system exceeds 2, calculating the multiple and partial spectral densities and coherences is mandatory. If this is not done, the results of analysis of the system will be both incomplete and incorrect.

The multiple coherent spectrum $s_{1:23}(f)$ is the part of the output process spectrum $s_{11}(f)$ generated by the linear dependence of the output upon both inputs. The partial coherent spectra $s_{12.3}(f)$ and $s_{13.2}(f)$ present the parts of the total output spectrum $s_{11}(f)$ generated by each input through the respective tract of linear transformation of each input into the output with account for possible correlation between the inputs. The random vector analogs of their integrals over the entire frequency band are the products of the output vector's variance by the squared multiple or partial cross-correlation coefficients.

The coherence functions can be regarded as frequency-dependent analogs of correlation coefficients in the case of random vectors. The multiple coherence $\gamma_{1:23}(f)$ shows the degree of linear interdependence between the output $x_{1,t}$ and the inputs $x_{2,t}, x_{3,t}$ with account for possible correlation between the inputs. Its analog for time-invariant vectors of random variables is the multiple cross-correlation coefficient. The partial coherence functions $\gamma_{12.3}(f)$ and $\gamma_{13.2}(f)$ characterize the linear dependence of the output process $x_{1,t}$ upon each of the inputs $x_{2,t}, x_{3,t}$ with account for possible linear relation between the inputs. They can be regarded as frequency-dependent analogs of partial cross-correlation coefficients between time-invariant random vectors.

The multiple and partial coherence functions define the Gelfand–Yaglom information rates for the case of multivariate random functions; they describe the information exchange between the output time series $x_{1,t}$ and both inputs $x_{2,t}, x_{3,t}$ and between the output time series $x_{1,t}$ with each of the inputs $x_{2,t}, x_{3,t}$ with account for possible linear dependence between the inputs. The expressions for respective information rates can be written as

- multiple information rate

$$i(x_{1,t}; x_{2,t}, x_{3,t}) = -\int_0^{f_N} \log[1 - \gamma_{1:23}^2(f)]\mathrm{d}f,$$

- partial information rates

$$i(x_{1,t}, x_{2,t}; x_{3,t}) = -\int_0^{f_N} \log[1 - \gamma_{12.3}^2(f)]\mathrm{d}f,$$

and

$$i(x_{1,t}, x_{3,t}; x_{2,t}) = -\int_0^{f_N} \log[1 - \gamma_{13.2}^2(f)]\mathrm{d}f.$$

These three equations present a generalization of the Gelfand–Yaglom information rate given for the bivariate case in their pioneering publication in 1957 and described in Chap. 7. Along with the coherence functions, the multiple and partial information rates serve as criteria of interdependences between the output and input time series with account for the dependence between the inputs. They can also serve as measures of causality and feedback criteria introduced by C. W. Granger for the time domain analysis (Chaps. 7 and 8).

The frequency response functions relating the output process to the inputs are complex-valued and can be regarded as frequency domain analogs of the partial regression coefficients. The moduli of these functions are the gain factors $g_{12.3}(f)$, $g_{13.2}(f)$, which show how each input process is transformed into a part of the output with account for possible dependence between the inputs. For example, if the value of the gain factor at some frequency is 0.5, it means that if the input process value at the same frequency is 1.0, the contribution of this input to the output will be 0.5.

The phase factors $\varphi_{12.3}(f)$ and $\varphi_{13.2}(f)$ show the delay between the input and output as functions of frequency. They are measured in radians and can be transformed into time delays $\tau_{1i.j}(f)$ by using the relation $\tau_{1i.j}(f) = \varphi_{1i.j}(f)/2\pi f$, $f > 0$. From physical considerations, the output should lag behind the inputs.

The equations defining all these functions are rather cumbersome, and the reader who wants to learn how to analyze multivariate systems should read the remarkable book by Bendat and Piersol (2010). The text given in this section is not sufficient for this purpose. The main goal of the examples given below is methodological because to the best of the author's knowledge examples of autoregressive analysis of multivariate ($M > 2$) time series in time and frequency domains do not exist in Earth and solar sciences and in any free or commercial software packages.

14.3 Analysis of a Simulated Trivariate Time Series

The primary goal of this example is to show how to investigate the behavior of an autoregressive stochastic system with two inputs and one output. The case with more than three inputs presents a simple extension of the trivariate case. All time series in the example are simulated and their length $T = N\Delta t = 1000$ time units (Fig. 14.1). A time series of such length is hardly realistic for annual climate observations, that is, for $\Delta t = 1$ year, but smaller sampling rates are used in climatology (e.g. Hashimoto et al. 2019) and in other Earth and solar sciences.

14.3.1 Time Domain Analysis

The trivariate time series in this chapter is an **AR**(2) sequence with the following AR coefficients:

$$\Phi_1 \approx \begin{bmatrix} 0.48 & 0.40 & 0.20 \\ -0.27 & -0.20 & -0.20 \\ -0.15 & 0.17 & 0.42 \end{bmatrix},$$

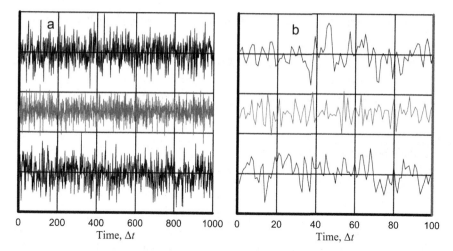

Fig. 14.1 Simulated trivariate time series: the entire set (**a**) and the first 100 values (**b**)

$$\mathbf{\Phi}_2 \approx \begin{bmatrix} -0.19 & 0.25 & -0.10 \\ -0.02 & -0.25 & 0.10 \\ -0.04 & -0.11 & 0.04 \end{bmatrix}.$$

This is equivalent to the following stochastic equation:

$$\begin{aligned}
x_{1,t} &= 0.48x_{1,t-1} + 0.40x_{2,t-1} + 0.20x_{3,t-1} - 0.19x_{1,t-2} \\
&\quad + 0.25x_{2,t-2} - 0.10x_{3,t-2} + a_{1,t} \\
x_{2,t} &= -0.27x_{1,t-1} - 0.20x_{2,t-1} - 0.20x_{3,t-1} - 0.02x_{1,t-2} \\
&\quad - 0.25x_{2,t-2} + 0.10x_{3,t-2} + a_{2,t} \\
x_{3,t} &= -0.15x_{1,t-1} + 0.17x_{2,t-1} + 0.42x_{3,t-1} - 0.04x_{1,t-2} \\
&\quad - 0.11x_{2,t-2} + 0.04x_{3,t-2} + a_{3,t}.
\end{aligned} \tag{14.3}$$

The true covariance matrix of the innovation sequence $\mathbf{a}_t = [a_{1,t}, a_{2,t}, a_{3,t}]'$ is

$$\mathbf{P_a} \approx \begin{bmatrix} 0.95 & 0.45 & 0.35 \\ 0.45 & 0.96 & 0.55 \\ 0.35 & 0.55 & 1.50 \end{bmatrix}. \tag{14.4}$$

The optimal model for the time series $\mathbf{x}_t = [x_{1,t}, x_{2,t}, x_{3,t}]'$ simulated in accordance with the above data was found to be an $\mathbf{AR}(2)$, and its parameters differ from the true ones within the limits allowed for the sampling variability at the confidence level 0.9. Specifically, the time domain model of simulated data is

$$x_{1,t} \approx 0.49x_{1,t-1} + 0.37x_{2,t-1} + 0.19x_{3,t-1} - 0.20x_{1,t-2} + 0.26x_{2,t-2}$$
$$- 0.13x_{3,t-2} + a_{1,t}$$
$$x_{2,t} \approx -0.27x_{1,t-1} - 0.18x_{2,t-1} - 0.23x_{3,t-1} - \mathbf{0.02}x_{1,t-2} - 0.21x_{2,t-2}$$
$$+ 0.10x_{3,t-2} + a_{2,t}$$
$$x_{3,t} \approx -0.18x_{1,t-1} + 0.15x_{2,t-1} + 0.45x_{3,t-1} - \mathbf{0.03}x_{1,t-2} - 0.10x_{2,t-2}$$
$$+ \mathbf{0.02}x_{3,t-2} + a_{3,t} \tag{14.5}$$

with the innovation covariance matrix

$$\mathbf{P_a} \approx \begin{bmatrix} 0.93 & 0.39 & 0.30 \\ 0.39 & 0.85 & 0.47 \\ 0.30 & 0.47 & 1.52 \end{bmatrix}$$

The bold font in Eq. (14.5) is used to show estimates that do not differ statistically from zero.

Both the true and estimated variances of the output component $x_{1,t}$ are 1.93 so that the variance of the first innovation sequence component is responsible for 48% of the total variance.

The correlation matrix of the innovation sequence is

$$\boldsymbol{\rho_a} \approx \begin{bmatrix} 1 & 0.44 & 0.25 \\ 0.44 & 1 & 0.42 \\ 0.25 & 0.42 & 1 \end{bmatrix}.$$

The correlation coefficients between the innovation components are statistically significant for $T = N\Delta t = 1000$ mutually independent variables.

The Granger causality for this trivariate time series is not strong: the error variances of forecasts of $x_{1,t}$ at the unit lead time $\tau = \Delta t$ within the framework of Kolmogorov–Wiener theory in the scalar and trivariate cases are 1.17 and 0.93, respectively, that is, the error variance of the trivariate forecast at the unit lead time is 20% smaller than the scalar one. Some improvements for the trivariate case exist for the input processes as well but they will not be discussed here.

Equation (14.5) describes a linear system with a closed feedback loop: the output is statistically dependent upon the inputs and vice versa.

A way to interpret this complicated model in the time domain is by determining the contributions of the input processes to the output. Multiplying the first equation of (14.5) by $x_{1,t}$ and finding mathematical expectations lead to the expression

$$\hat{\sigma}_1^2 = \sum_{k=1}^{2} [\varphi_{11}^{(k)} R_{11}(k) + \varphi_{12}^{(k)} R_{21}(k) + \varphi_{13}^{(k)} R_{31}(k)] + P_{11}, \tag{14.6}$$

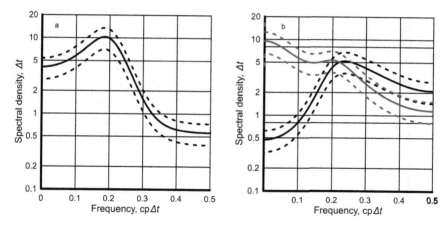

Fig. 14.2 Spectral densities of time series components: outputs (**a**) and inputs (**b**)

which describes the contributions of the output's and inputs' past values to the output variance $\sigma_1^2 \approx 1.93$; here, $R_{ij}(k)$ are the elements of the covariance matrix $\mathbf{R}(k)$. We will not discuss it further.

As the time series dimension and the **AR** order p increases, the interpretation of the time domain model gets more and more cumbersome. Therefore, the frequency domain information becomes especially important because it allows one to study relations between the outputs and the inputs with account for possible correlation between the inputs in a relatively simple and easily understandable manner.

14.3.2 Frequency Domain Analysis

The spectral densities $s_{11}(f)$, $s_{22}(f)$, $s_{33}(f)$ of the time series components $x_{1,t}$, $x_{2,t}$ and $x_{3,t}$ shown in Fig. 14.2 are not monotonic: the output spectrum has a clear maximum at about 0.18 cpΔt, and the first input is peaked at about 0.22 cpΔt. The second input has a very smooth spectrum decreasing with frequency with a hump at about 0.20 cpΔt, similar to the characteristic behavior of the global annual temperature and other climatic processes (Privalsky and Yushkov 2018). The spectra are smooth but not monotonic. As usual, the confidence bounds are shown at the confidence level 0.90, and they are approximate. Such spectra could belong to climatic or other geophysical processes.

The degree of linear dependence between the output and inputs as a function of frequency can be seen in Fig. 14.3. The multiple coherence stays significant within the entire frequency band, and it exceeds 0.6 in the band from 0.1 to 0.3 cpΔt, with the maximum value of 0.89 at about 0.2 cpΔt. This means that the multiple coherent spectrum $s_{1:23}(f)$ will amount to almost 80% of the total spectral density $s_{11}(f)$ at $f = 0.2$ cpΔt and will stay above 40% of $s_{11}(f)$ inside the frequency band from

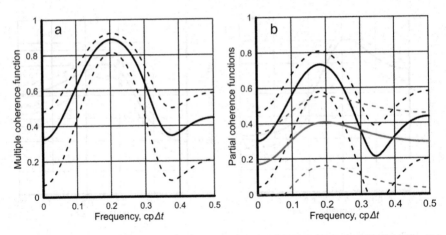

Fig. 14.3 Coherence functions between the output and inputs: multiple $\gamma_{1:23}(f)$ (**a**) and partial $\gamma_{12.3}(f), \gamma_{13.2}(f)$, black and gray (**b**)

0.1 to 0.3 cpΔt. This linear dependence of the output upon inputs is strong within a wide frequency band.

The partial coherence function $\gamma_{12.3}(f)$ shown in Fig. 14.3b behaves similar to the multiple coherence but its values are smaller. Its maximum is 0.73 at 0.18 cpΔt, and it explains about 50% of the output spectral density at that frequency. The output is less dependent upon the second input but the values of partial coherence $\gamma_{13.2}(f)$ stay significant everywhere above 0.1 cpΔt.

The coherent spectra—the multiple spectrum $s_{1:23}(f)$ and the partial spectra $s_{12.3}(f)$ and $s_{13.2}(f)$ behave in accordance with the behavior of respective coherences and spectra. As seen from Fig. 14.4, the multiple coherent spectrum $s_{1:23}(f)$

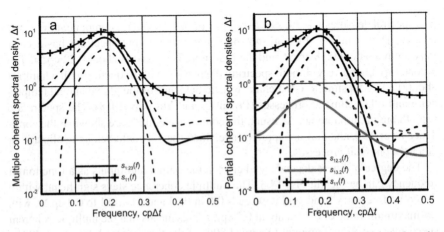

Fig. 14.4 Multiple (**a**) and partial coherent spectral densities (**b**). The lines with symbols show the output spectrum.

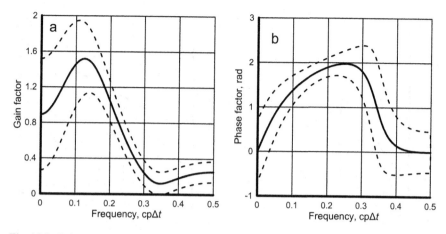

Fig. 14.5 Gain (**a**) and phase (**b**) factors of the frequency response function connecting the output $x_{1,t}$ to the first input $x_{2,t}$

is quite close to the spectrum $s_{11}(f)$ at intermediate frequencies; the partial coherent spectrum $s_{12.3}(f)$ behaves in the same manner. The coherent spectrum $s_{13.2}(f)$ is smaller by an order of magnitude, and its lower confidence limit goes to zero and cannot be shown in Fig. 14.4.

The behavior of each frequency response function that transforms the inputs into the output is convenient to analyze separately, showing the gain and phase factors in one figure. The gain factor $g_{12.3}(f)$ shown in Fig. 14.5 demonstrates that respective frequency response function amplifies variations of the first input at frequencies up to 0.2 cycles per Δt and suppresses them at higher frequencies. The phase factor $\phi_{12.3}(f)$ changes in a wide range from 0 rad to 2 rad and then goes back to zero. As mentioned before, the phase shift $\phi(f)$ can be recalculated into the time shift $\tau(f)$; for example, at $f = 0.25$ cpΔt, the phase $\phi_{12.3}(0.25) \approx 2$ rad so that the time shift $\tau(f) \approx 1.3\ \Delta t$.

The frequency response function connecting the output to the second input generally suppresses variations of $x_{3,t}$ when transforming it into the output $x_{1,t}$ (Fig. 14.6). The strictly linear change of the phase factor $\phi_{13.2}(f)$ with frequency means that the time shift between the input and output is constant. In this case, $\tau(f) = \Delta t$ at all frequencies. This is an intentional effect of the simulation procedure designed to detect the time shift of one sampling interval between $x_{1,t}$ and $x_{3,t}$.

Thus, the frequency domain analysis provides information about this or any other stationary multivariate time series which can hardly be obtained in any other way. In addition to what can be learned about bivariate time series in the time domain, it includes multiple and partial coherence functions and spectra. The gain and phase factors are obtained for each tract of the system: from $x_{2,t}$ to $x_{1,t}$ and from $x_{3,t}$ to $x_{1,t}$. All these functions are calculated with account for possible interdependence between the input processes.

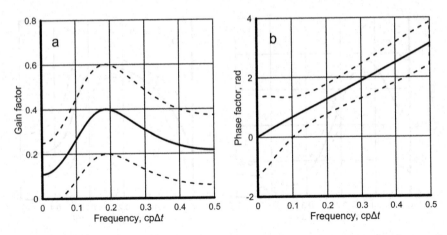

Fig. 14.6 Gain (**a**) and phase (**b**) factors of the frequency response function connecting the second input $x_{2,t}$ to the output $x_{1,t}$

As in the bivariate case, this information illustrates physical properties of the time series and, along with the results of time domain analysis, can be taken into account when building physical models of geophysical processes.

The practical application of multivariate parametric analysis at climatic time scales can be limited by the lack of sufficient data but other areas such as atmospheric sciences, oceanography, and other geosciences can benefit from this powerful mathematical tool. Publications containing multivariate ($M > 2$) autoregressive time series analysis in frequency domain seem to be nonexistent in Earth and solar science but an example of multivariate nonparametric spectral analysis can be found in Sokolova et al. (1992). This concludes the example with the trivariate time series simulated in accordance with Eqs. (14.3) and (14.4).

The software for frequency domain analysis of M-variate time series with $M > 2$ is given in R in Wei (2019, Sect. 9.4.4 and pp. 358–361). In his Example 9.3(d), the author studies a 5-variate time series of length $N = 90$ and selects the AR order $p = 4$ for it. This means that the number of AR coefficients which need to be estimated is $M^2 p = 100$ while the total number of terms is 450 and the scalar components are strongly interdependent, according to the coherence estimates given in the example. As the number of coefficients to be estimated in this example is comparable to the total number of data, the estimates of autoregressive coefficients and all functions of frequency are statistically unreliable. Moreover, the results of analysis contain only ordinary coherence functions within the ten pairs of scalar components belonging to that 5-variate time series. The multiple and partial coherent spectra and coherence functions as well as gain and phase factors are not calculated, and the results do not contain any information about confidence limits for all estimated quantities. At the same time, this seems to be a rare case of obtaining frequency domain characteristics of a stationary multivariate time series through its autoregressive model.

14.4 Analysis of Climatic Time Series

At the sampling rate $\Delta t = 1$ year, it is not easy to find a multivariate time series whose length would be sufficient for applying multivariate autoregressive analysis to it. Thus, with $M = 3$ and $N = 150$, the necessity to estimate $9p$ coefficients may limit the AR order p to one due to the influence of order selection criteria. Having in mind the methodological goals of this book, the three-dimensional climatic time series in the example below are based upon the monthly surface temperature data provided by the University of East Anglia at their Web site https://crudata.uea.ac.uk/cru/data/temperature/ (also see Morice et al. 2012). The Nyquist frequency f_N for such data is 0.5 cycles per month, or 6 cpy. The task here is to study the internal structure of two linear systems describing the dependence of the global temperature (GLOBE) upon the hemispheric temperature (NH, SH) and upon the ocean and land data (OCEAN and LAND). The full length of the time series is 170 years (2040 months) but we will also study the time series over the interval from 1850 through 1919 (840 months) and from 1920 through 2019 (1200 months). The full original initial data set is shown in Fig. 14.7. As seen from the figure, the entire time series can hardly be treated as stationary while the two shorter data sets allow one to regard the data as samples of stationary random processes. Yet, the full-time series from 1850 through 2019 will also be analyzed as an additional methodical example. Altogether, there will be six examples of trivariate time series analyses: three for the interval from 1850 through 2019 and three for the shorter intervals from 1850 through 1919 and from 1920 through 2019. The time series that begin in 1920 contain relatively reliable data studied in particular in Dobrovolski (2000) and in Chap. 5.

Consider first the system with the GLOBE temperature as the output and the hemispheric temperature time series NH and SH as inputs. The optimal models selected for the sets 1850–2019 and 1920–2019 were **AR**(2); the model for the

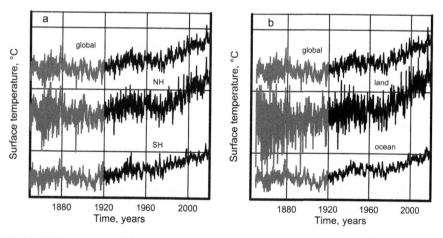

Fig. 14.7 Global and hemispheric temperature (**a**) and global, ocean, and land temperature (**b**). The 1850–2019 and 1920–2019 data are shown in black and black

time series from 1850 through 1919 was **AR**(1). The global temperature in the file
GLOBE presents an average of the hemispheric temperatures NH and SH, that is,
the GLOBE time series presents a strictly linear function of the inputs NH and SH.
Consequently, all three coherent functions—the multiple coherence $\gamma_{1:23}(f)$ and the
partial coherences $\gamma_{12.3}(f)$ and $\gamma_{13.2}(f)$ must be equal to 1 at all frequencies, and
this is what happens with the actual estimates for all three cases with the hemispheric
time series as inputs. With the coherence functions equal to one, the systems become
unstable in the sense that the variances of estimated AR coefficients may not be
reliable. This behavior of the data is absolutely correct, and we will continue to
discuss the frequency domain properties of the models.

The estimated spectra of the global and hemispheric time series are monotonic,
and their dynamic range amounts to three orders of magnitude (Fig. 14.8). The results
for the two shorter time series with the hemispheric temperature time series as inputs
are similar to what we have for the entire time series.

As the coherence functions are equal to one, all three coherent spectra—the
multiple spectrum $s_{1:23}(f)$ and the partial spectra $s_{12.3}(f)$ and $s_{13.2}(f)$ coincide
with the respective spectra as shown in Fig. 14.8.

According to the data for the entire globe and for the hemispheres, the contributions of hemispheres to the global temperature are identical so that the gain factors
$g_{12.3}(f)$ and $g_{13.2}(f)$ (not shown) are equal to 0.5 at all frequencies for all three
versions. The response of the global temperature to variations of hemispheric temperatures is immediate: the temperature variations occur without any phase shift or time
lag at all frequencies. These data obviously agree with each other.

On the whole, this analysis did not bring any unexpected results: the behavior of the
output process (the global temperature) reflects the method of determining the global
temperature as the average of hemispheric temperatures. This rather complicated way
of studying the global temperature used just above has brought us to the results that

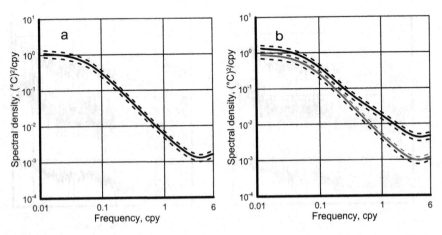

Fig. 14.8 Spectra of global (**a**) and hemispheric (**b**) monthly temperature (black and gray for the
northern and southern hemispheres), 1850–2019

could have been obtained through a linear regression between the global temperature as a function of hemispheric temperature without any errors in the values of regression coefficients. Variations between estimates of both gain factors and the regression coefficients occur only in the fourth decimal digit. The gain factors and the regression coefficients are always equal to 0.500.

Consider now the global data for different time intervals as functions of the areas taken on the globe by the oceans and land, that is, consider the dependence of the global temperature upon the oceanic and terrestrial temperature (Fig. 14.7b). The optimal models selected by the order selection criteria were **AR**(2) in all three cases. The spectral densities of oceanic and terrestrial temperature behave in the same manner as hemispheric data, and the shapes of spectral densities do not differ much from what is shown in Fig. 14.8. However, analysis shows that a strictly linear relation to the global temperature has disappeared and that coherence functions are not equal to unity any more. The values of all three coherence functions are still quite high (Fig. 14.9) but the partial coherence $\gamma_{12.3}(f)$ between the global and oceanic data decreases to about 0.8 at higher frequencies. This deviation from a linear system is not very important because the coherence stays sufficiently high at climatic frequencies (below 0.5 cpy). Yet, the loss of linearity means the lack of complete correspondence between the three time series that is difficult to explain.

According to the results of analysis of the entire time series, the gain factors $g_{12.3}(f)$ and $g_{13.2}(f)$ practically coincide with the relative areas of oceanic and terrestrial surfaces: 0.71 and 0.29, that is, 71% and 29% for the ocean and land areas.

However, the results for the time intervals from 1920 through 2019 differ from the results for the entire time series in one important point: the gain factors $g_{12.3}(f)$ and $g_{13.2}(f)$ remained almost constants at all frequencies but their absolute values changed from 0.71 and 0.29 to about 0.63 and 0.35 at climatic frequencies (Fig. 14.10). This change from the results for the interval from 1850 is statistically

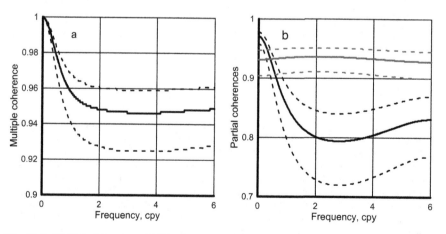

Fig. 14.9 Multiple (**a**) and partial (**b**) coherence functions between the global temperature and the oceanic (black) and terrestrial (gray) temperature, 1850–2019

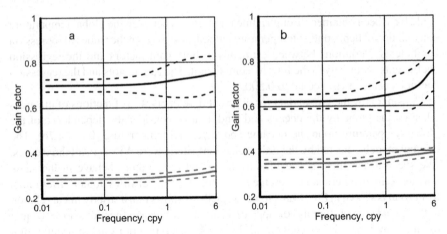

Fig. 14.10 Gain factors $g_{12.3}(f)$ and $g_{13.2}(f)$ for the time series beginning in 1850 (**a**) and in 1920 (**b**)

significant. The results for the shortest time series from 1850 through 1919 also look improper: the gain factors between the global temperature at the output and the oceanic and terrestrial data as inputs amount to approximately 0.75 and 0.25. The cause of this nonhomogeneity in the properties of time series of global, oceanic, and terrestrial monthly temperature is unclear. This difference in statistical characteristics obtained from the time series starting in 1850 and in 1920 may be the reason for the change in predictability of the three time series containing data for the southern hemisphere mentioned in Chap. 5. Similar results can be obtained if analysis is conducted within the framework of mathematical statistics, through multiple linear regression.

Obviously, the results for the two shorter time series cannot be correct because the areas taken by oceans and land are the same—71% and 29%—for all three cases. This odd behavior of the time series of oceanic and terrestrial time series as a function of time needs to be explained.

References

Bendat J, Piersol A (2010) Random data. Analysis and measurements procedures, 4th edn. Wiley, Hoboken

Dobrovolski S (2000) Stochastic climate theory. Models and applications. Springer, Berlin

Gelfand I, Yaglom A (1957) Calculation of the amount of information about a random function contained in another such function. Uspekhi Matematicheskikh Nauk 12:3–52. English translation: American Mathematical Society Translation Series 2(12):199–246, 1959

Hashimoto H, Wang W et al (2019) High-resolution mapping of daily climate variables by aggregating multiple spatial data sets with the random forest algorithm over the conterminous United States. Int J Climatol http://doi/14.1002/joc.5995

Morice C, Kennedy J, Rayner N, Jones D (2012) Quantifying uncertainties in global and regional temperature change using an ensemble of observational estimates: The HadCRUT4 data set. J Geophys Res Atmospheres 137:D08101–D08122

Privalsky V, Yushkov V (2018) Getting it right matters: climate spectra and their estimation. Pure Appl Geoph 175:3085–3096

Shumway R, Stoffer D (2017) Time series analysis and its applications. Springer, Heidelberg

Sokolova S, Rabonovich A, Chu K (1992) On the atmosphere-induced sea level variations along the western coast of the Sea of Japan. La Mer 30:191–212

Wei W (2019) Multivariate time series analysis and applications. Wiley, Hoboken

Chapter 15
Summary and Recommendations

Abstract Climatology and related geophysical and solar sciences have been going for decades through a crisis in research involving mathematical statistics and time series analysis. Mathematical statistics is improperly applied in studies that require the use of time series analysis (exploring multivariate time series in time and frequency domains) and not applied where it is necessary (probability density functions, confidence bounds for estimated statistical characteristics with account for correlation structure of the time series). Time series are erroneously treated as random vectors though random vectors do not have correlation functions and spectra, PDFs are rarely analyzed, reliability of estimates is assessed without taking into account serial correlation within the time series, estimates of statistics are given without confidence bounds, an incorrect test is applied to assess significance of spectral peaks. Studies of teleconnections are based upon improper estimates of cross-correlation coefficients while time series reconstructions, first of all, reconstructions of climate, use the cross-correlation coefficients and regression equations which are not applicable to time series. The classical theory of extrapolation by Kolmogorov and Wiener is not known. This book provides many examples, and this chapter sums up practical recommendations helping to properly analyze and forecast scalar and multivariate time series.

The branches of Earth and solar sciences related to time series analysis and forecasting are experiencing a crisis. It is caused by unawareness of methods developed many decades ago for those purposes within the frameworks of theory of random processes and information theory. This is happening in disregard of many monographs on time series analysis written by professionals for applied sciences and engineering (e.g., Blackman and Tukey 1958; Yaglom 1962; Bendat and Piersol 1966; Jenkins and Watts 1968; Box and Jenkins 1970; Shumway and Stoffer 1999). Some of these books have been updated and republished up to four (Bendat and Piersol 2010; Shumway and Stoffer 2017) or even five times (Box et al. 2015), but the methods of time series analysis and extrapolation described in those books still remain poorly known in Earth and solar sciences. To some extent, it is also true for proficient books written, in particular, for applications in climatology and atmospheric sciences such as von Storch and Zwiers (1999) and Wilks (2011). The crisis

© Springer Nature Switzerland AG 2021
V. Privalsky, *Time Series Analysis in Climatology and Related Sciences*,
Progress in Geophysics, https://doi.org/10.1007/978-3-030-58055-1_15

involves both scalar and multivariate time series tasks including spectral analysis of scalar time series, statistical forecasting, analysis of scalar and multivariate time series in time and frequency domains, time series reconstruction, and issues related to statistical reliability of results.

If the term time series is used for any data that is supposed to be analyzed, predicted, or reconstructed, the researcher must avoid methods that do not recognize and do not use the property of data's dependence upon the time argument. Specifically, this requirement automatically excludes the classical mathematical statistical quantities such as cross-correlation coefficients, scalar of multivariate regression equations, and whatever else that does not take into account the time or frequency arguments.

Theory of random processes with time series analysis as a part of it and with information theory contributions to it have been in existence for decades and methods of analysis of time-dependent random data started to be applied for solving engineering and scientific problems since long ago. Therefore, if a method of time series forecasting not based upon the theory of stationary random processes is used, one must show that its theoretical basis is at least as good as in methods of extrapolation developed by N. Wiener (1949), A. Yaglom (1962), and G. Box and G. Jenkins (1970). New methods of scalar and multivariate spectral analysis should also have a solid theoretical foundation that would justify their introduction on a par with the methods developed by R. Blackman and J. Tukey (1958), D. Welch (1967), J. Burg (1967), G. Jenkins and D. Watts (1968), D. Thomson (1982), Bendat and Piersol (2010), and with the multivariate autoregressive spectral analysis introduced in this book. Incidentally, a Monte Carlo experiment cannot serve as a replacement or substitute for a mathematical proof.

Working with scalar time series should include verification of stationarity and ergodicity assumptions (unless the stationarity is obvious visually), estimation of spectral density and, if necessary, the statistical predictability. In many cases, none of these subjects gets attention in climatology and other Earth and solar sciences. At the same time, examples of disregard for proper and easily accessible methods of time series analysis including spectral estimation are quite common. Mathematical statistics is often incorrectly used to determine the significance of results of spectral analysis and multivariate time domain research.

In the multivariate case, the approach to analysis is incorrect in most cases, because it is traditionally based upon regression analysis—a method of classical mathematical statistics which is inapplicable to random functions of time, that is, to time series. This happens in research related to spectral analysis in general, to detecting and studying teleconnections within different Earth and solar systems and to time series reconstructions. Climate reconstructions conducted in this manner are used for supporting the anthropogenic global warming hypothesis and thus play a confusing role. Such quantities as multiple and partial coherence functions and coherent spectra along with gain and phase factors used in different branches of engineering are practically unknown in Earth and solar sciences.

The two major goals of the book were to describe methods that should be used for analysis and forecasting of scalar and multivariate (mostly, bivariate) time series and to demonstrate that methods based upon theory of random processes provide much more useful information about geophysical and solar systems than what can be obtained with traditional methods. It has to be stressed that methods of analysis of scalar and multivariate time series used in this book have been in existence for many decades and practically all of them are easily accessible in free and commercial software.

A major point in this book is that the autoregressive approach presents a proper way to achieve a reliable solution for the task of analysis and prediction of stationary time series, both scalar and multivariate. The only requirements here are that the time series is stationary and that it is sufficiently long for being analyzed. This statement applies to climatology, Earth sciences in general, and to related solar sciences. The autoregressive approach is especially useful for analysis of relatively short time series. If the time series is long, that is, if its length exceeds the longest time scale of interest by at least an order of magnitude, the properly applied methods of nonparametric analysis also ensure reliable estimates of the spectra, spectral matrices, and all other frequency-dependent characteristics of time series. In such cases, the autoregressive and nonparametric methods are equally useful as means of spectral estimation. The important information which is lost if the nonparametric approach is used includes the explicit scalar or multivariate stochastic difference equations—a ready-to-apply tool for linear extrapolation, for determining the dependence of the time series upon its past and upon other components of multivariate time series, for Granger causality and feedback research, and for other topics of interest. Results of time domain analysis in the form of stochastic difference equations may give physicists some ideas about describing what is happening in a geophysical or solar system using the tools of fluid dynamics.

A vital goal of the book was to show how geophysical and solar time series should be analyzed through autoregressive modeling within the framework of theory of random processes, information theory, and mathematical statistics. Some of the previously published monographs include a lot of useful information on these subjects but contain only a few or no examples of practical analysis, especially in the multivariate case. The parametric multivariate analysis of real data in time and frequency domains does not seem to be known in Earth sciences. The few exceptions for the bivariate case belong to this author (Privalsky 1988, 2015, 2018; Privalsky and Jensen 1995) and obviously do not make a difference. The book includes a rather large number of practical examples intended to draw attention of researchers working with time series in climatology, geophysics in general, and in solar sciences.

The autoregressive modeling presents an efficient tool for studying scalar and multivariate time series. It is based upon a sound mathematical theory and allows one to describe time series with stochastic difference equations in the time domain and then to use the equations to estimate the frequency domain properties expressed with spectral densities, coherence functions, coherence spectra, and frequency response functions including frequency-dependent time lags. As mentioned before, it provides an immediate way to extrapolate (forecast, predict) the time series and determine the variance of extrapolation error.

In the scalar case, the advantages of autoregressive (or maximum entropy) spectral analysis are shown and illustrated with many examples, mostly at climatic time scales but also at a monthly sampling rate. One time series—a Madden–Julian Oscillation data set—has been studied at the daily sampling rate. The examples show that spectra of most geophysical processes are smooth and, with the exclusion of Quasi-Biennial Oscillation, do not contain significant cyclicity at time scales measured in years. This result is based upon mathematically correct analysis of many scalar time series, and it disagrees with numerous publications in climatology and related sciences where "statistically significant spectral peaks are found" at many frequencies due to improper applications of spectral analysis.

Climatic and many other time series are short and efficient approaches to estimating their spectra is through autoregressive models (maximum entropy spectral analysis) and through Thomson's multitaper method (MTM). If the spectrum is not too complicated, the autoregressive approach provides useful information about the behavior of the time series in the time domain in the form of low-order stochastic difference equations. The autoregressive approach is also efficient with time series that have complicated spectra, but in all cases, the order of autoregression must be determined with order selection criteria developed in information theory. The nonparametric spectral estimates through MTM agree with the AR estimates.

The issue of statistical predictability is rarely discussed and different methods of forecasting are applied in climatology and other Earth sciences without even mentioning the classical Kolmogorov–Wiener theory of extrapolation (KWT) developed about 80 years ago for stationary random processes. This omission is regrettable because KWT had been built in a way that ensures the least possible variance of extrapolation error in the linear case and the absolutely minimal error variance in the Gaussian case. The examples of KWT forecasts of scalar geophysical time series given in the book illustrate the ability of the theory and also show some non-Gaussian cases when KWT may not be efficient. The theory covers the extrapolation of multivariate processes, and a bivariate extrapolation case is briefly discussed in Chap. 8.

A detailed knowledge of information theory is hardly required for analyzing a multivariate time series, but one has to understand that multivariate time series should not be treated as time-invariant random vectors and that information about relations between time series cannot be measured with cross-correlation coefficients. The knowledge of frequency-dependent characteristics including spectra and coherent spectra, coherence functions, gain and phase factors is necessary for describing relations between scalar components of multivariate time series.

The examples of time and frequency domain analysis are given in the book as illustrations of the high productivity that can be achieved when proper methods are applied to bivariate time series. Examples of analysis of trivariate time series are given to allow the reader to see what additional information can be obtained for time series containing more than two scalar components. An important goal of the examples of teleconnection analysis and time series reconstruction given in the book is to demonstrate that time series must be treated as random functions of time rather than as time-invariant random vectors. The results of such analysis are shown to contain detailed information about teleconnection systems in both time and frequency domains which cannot be received through traditional regression analysis.

It is up to the reader to decide whether the author's attempt to show the advantages of applying proper methods has been successful or unsuccessful, but the reader is strongly advised to avoid the following errors in time series analysis occurring in climatology, other Earth sciences (with the exception of solid Earth physics), and in solar research.

1. Do not ignore methods developed for time series analysis on the basis of theory of random processes, information theory, and mathematical statistics. The recommended mathematical literature is Percival and Walden (1993), Bendat and Piersol (2010), Box et al. (2015), Thomson and Emery (2014), Shumway and Stoffer (2017). The book by Yaglom (1962) is necessary for anyone who is involved in time series forecasting.
2. Do not use estimates of statistical parameters and functions without respective confidence intervals at a specified confidence level, which are always provided within the methods developed by professionals. Estimates without confidence bounds have absolutely no value.
3. Do not determine the significance of spectral peaks by comparing your spectral estimate with the upper confidence limit of the spectrum of an equivalent Markov chain. If your spectrum estimate lies above the limit, it means that the time series has not been generated by a Markov process. It has nothing to do with the statistical significance of the peak.
4. Do not forget to estimate the probability density function of time series, both scalar and multivariate. If it is Gaussian (normal), the results of analysis become more general than in non-Gaussian cases.
5. Do not filter your time series without first proving the necessity of the operation. Filtering causes drastic changes in all statistical characteristics and in confidence bounds for their estimates; it also changes the shape of the spectrum. Parametric methods of spectral analysis allow one to avoid interaction between high-frequency "noise" and lower frequency "signal." In geophysics, processes whose spectra increase with frequency seem to be nonexistent.
6. Do not use any linear method of time series extrapolation (prediction, forecasting) not based upon the Kolmogorov–Wiener theory without having proved that the method agrees with it. The theory ensures the smallest error variance of linear extrapolation for any time series generated by a stationary random process. If the process is Gaussian, using KWT ensures the smallest possible error variance of forecasting, linear or nonlinear. Nonlinear methods of extrapolation should be applied only to non-Gaussian processes.
7. Do not use nonparametric methods of spectral analysis (e.g., Welch's and Bendat and Piersol's) without making sure that the time series is long enough for obtaining statistically reliable results. Exception: Thomson's multitaper method designed, in particular, for short time series that may contain communication signals.

8. Do not estimate too many statistical parameters when analyzing a time series. The number of such estimates must be at least an order of magnitude smaller than the time series length. Estimates obtained in serious violation of this rule are useless.
9. Do not determine the order of parametric models of time series without the help of order selection criteria.
10. Do not apply methods of classical mathematical statistics to describe relation between time series. Those relations are frequency dependent and must be studied in both time and frequency domains using methods of time series analysis.
11. Do not use the cross-correlation coefficient or any other single value of cross-correlation function to characterize interdependence between time series. It is described with the coherence functions, coherent spectra, and gain and phase factors.
12. Do not use regression equations to calculate values of a time series as a function of other time series. In particular, do not use a regression equation connecting time series to each other to reconstruct past values of a time series over the time interval when it has not been known. The cross-correlation coefficient and regression equation cannot describe relation between time series.
13. Finally, do not ever evade the stage of time series analysis in the frequency domain; the knowledge of spectral densities and other functions of frequency is vital for understanding the nature and properties of any stationary time series.

Ignoring the principles of time series analysis developed within the framework of theory of random processes and information theory is wrong and leads to incorrect conclusions about probabilistic and physical properties of scalar and multivariate time series. It brings confusion to our ability to understand physical processes occurring on the Earth and on the Sun, including the nature of the current global warming.

References

Bendat J, Piersol A (1966) Measurement and analysis of random data. Wiley, New York
Bendat J, Piersol A (2010) Random data, 4th edn. Wiley, Hoboken
Blackman R, Tukey J (1958) The measurements of power spectra. Dover Publications, New York
Box G, Jenkins M (1970) Time series analysis. Forecasting and control. Wiley, Hoboken
Box G, Jenkins G, Reinsel G, Liung G (2015) Time series analysis. Forecasting and control, 5th edn. Wiley, Hoboken
Burg J (1967) Maximum entropy spectral analysis. Paper presented at the 37th Meeting of Society of Exploration Geophysicists, Oklahoma City, OK, October 31, 5 pp
Jenkins G, Watts D (1968) Spectral analysis and its applications. Holden Day, San Francisco
Percival D, Walden A (1993). Spectral analysis for physical applications. Cambridge University Press
Privalsky V (1988) Stochastic models and spectra of interannual variability of mean annual sea surface temperature in the North Atlantic. Dyn Atmos Ocean 12:1–18
Privalsky V (2015) On studying relations between time series in climatology. Earth Syst Dyn 6:389–398

Privalsky V (2018) A new method for reconstruction of solar irradiance. JASTP 172:138–142

Privalsky V, Jensen D (1995) Assessment of the influence of ENSO on annual global air temperature. Dyn Atmos Ocean 22:161–178

Privalsky V, Yushkov V (2018) Getting it right matters: climate spectra and their estimation. Pure Appl Geoph 175:3085–3096

Shumway R, Stoffer D (1999) Time series analysis and its applications. Springer, Heidelberg

Shumway R, Stoffer D (2017) Time series analysis and its applications, 4th edn. Springer, Heidelberg

von Storch H, Zwiers F (1999) Statistical analysis in climate research. Cambridge University Press, Cambridge

Thomson D (1982) Spectrum estimation and harmonic analysis, P. IEEE 70:1055–1096

Thomson R, Emery W (2014) Data analysis methods in physical oceanography, 3rd edn. Elsevier, Amsterdam

Welch P (1967) The use of Fast Fourier Transform for the estimation of power spectra: A method based on time averaging over short, modified periodograms. IEEE Trans Audio and Electroacoustics, AU-15, pp 70–73. https://doi.org/10.1109/tau.1967.1161901

Wiener N (1949) Extrapolation, interpolation, and smoothing of stationary time series, with engineering applications. Wiley, New York

Wilks D (2011) Statistical methods in atmospheric sciences, 3rd edn. Academic Press, Oxford

Yaglom A (1962) An introduction to stationary random functions. Prentice Hall, Englewood Cliffs

Printed in the United States
By Bookmasters